Light Detectors, Photoreceptors, and Imaging Systems in Nature

Light Detectors, Photoreceptors, and Imaging Systems in Nature

JEROME J. WOLKEN

Carnegie Mellon University
Biological Sciences
Pittsburgh, Pennsylvania

New York   Oxford
OXFORD UNIVERSITY PRESS
1995

Oxford University Press

Oxford New York Toronto
Delhi Bombay Calcutta Madras Karachi
Kuala Lumpur Singapore Hong Kong Tokyo
Nairobi Dar es Salaam Cape Town
Melbourne Auckland Madrid

and associated companies in
Berlin Ibadan

Published by Oxford University Press, Inc.,
200 Madison Avenue, New York, New York 10016

Oxford is a registered trademark of Oxford University Press

Library of Congress Cataloging-in-Publication Data
Wolken, Jerome J. (Jerome Jay), 1917–
Light detectors, photoreceptors,
and imaging systems in nature/
Jerome J. Wolken.
p. cm. Includes bibliographical references and index.
ISBN 0-19-505002-9
1. Photoreceptors. 2. Visual pigments.
3. Photobiology.
I. Title.
QP481.W58 1994 591.1'823—dc20 94-14160

9 8 7 6 5 4 3 2 1

Printed in the United States of America
on acid-free paper

Preface

In *Light Detectors, Photoreceptors, and Imaging Systems in Nature*, I explore some of the many ways light is intimately linked with life. Light is necessary for photosynthesis and vision, as well as the photobehavior of plants and animals. The reception of light is important to our health and well-being. Various wavelengths of light are being used as a tool for diagnosing diseases and to restore health.

Therefore, I begin with the physical properties of light and the electromagnetic spectrum of energies, including the solar spectrum that reaches the Earth, which spans the ultraviolet, the visible, and the infrared—wavelengths from near 200 nm to about 950 nm. The effective energies for biological photoprocesses are generally limited to the visible wavelengths from around 340 nm to 780 nm. Many of the photobiological phenomena that are discussed in this book occur within these wavelengths and center around 500 nm, the solar energy peak in the blue-green. Living organisms utilize solar energy via their pigment molecules (e.g., carotenoids, chlorophyll, flavins, phytochromes, retinals), which are chemically structured to absorb the visible wavelengths of light. The biosynthesis of pigment molecules, their chemical structure, and absorption spectra are described.

In reviewing the various photoreceptor systems in nature, I first consider the process of photosynthesis and how chlorophyll in the chloroplast is molecularly structured to transduce the light energy absorbed to chemical energy in the photoprocess. I then investigate the photoreceptor systems of unicellular organisms (algae, fungi, bacteria) that respond to light by oriented bending, *phototropism*, or that freely move about, *phototaxis*. Organisms that illustrate the phototactic phenomena include the fungus *Phycomyces* and the flagellated algae, *Euglena* and *Chlamydomonas*, as well as the bacterium *Halobacterium halobium*. The photoreceptors of all these organisms reside in the cell membrane or in structures that are highly ordered membranes, *crystalline* structures, in which the photosensitive pigment is associated.

Assuming that similar structures in other organisms have evolved into photosensory cells, I hypothesize that these cells developed a lens through which to focus the light on their photoreceptors and became a simple imaging eye. A comparative structural analysis of invertebrate and vertebrate eyes follows. From evolutionary considerations, I work backwards, beginning with the most highly evolved, the vertebrate eye, showing how it is structured to function for vision. I then consider how bird and fish eyes differ from those of land animals in adapting to their environments. In doing so, I compare their optical systems, retinal photoreceptors, visual pigments, absorption spectra, spectral sensitivity, and color vision.

The most interesting eyes and visual systems are found in the invertebrate arthropods (insects, crustaceans, arachnids) and in molluscs. Every conceivable device for forming an image is found among them, from pinhole eyes, to simple camera eyes, to compound eyes, to eyes with refracting optics. Their lenses vary in shape from spherical to aspherical and have graded indices of refraction. Some eyes have reflecting surfaces, mirrors, and fiberoptic light guides, all to improve their ability to see.

Photoreceptors are not restricted to animal eyes, for blinded and eyeless animals sense light. This is possible because photodetectors, extraocular photoreceptors, are found over regions of the skin, in deeper tissue neural cells, and in the pineal organ of the brain. The reception of light by these photodetectors greatly affects animal behavior.

For example, specific photodetectors function as a light meter in measuring light intensity and in clocking the hours of light and darkness. The timing of these photoperiods (light-dark) is related to circadian rhythms. Light reception via extraocular photoreceptors influences the synthesis of hormones and determines the timing of reproductive sexual cycles. Extraocular photoreceptors can function either alone, with the neuroendrocrine system and/or with the visual system. Not surprisingly, we will find later in this book that the extraocular photoreceptors have similar molecular structure as all photoreceptors (chloroplasts, retinal rods, and cones).

In summarizing published studies of photoreceptors and visual systems in nature, I indicate how knowledge about them can be exploited for the development of experimental photochemical systems for converting light energy to chemical and electrical energy; the creation of a photochemical system analogous to a computer for receiving, storing, and transmitting information; the manufacture of devices to improve solar energy collection; and the realization of new imaging lenses for the visually impaired. Such experimental systems and model devices may enhance human lives, as well as our technology.

This book is a personal account and does not cover all the research advances in photobiological and visual science. The organisms, plants, and animals that I selected for investigating photoreceptor phenomena are primarily those I have studied and whose photoreceptor, optical, and visual systems have fascinated me over many years. The topics covered are highly specialized, and the mechanisms that underlie photobiological phenomena are highly complex. Major technological advances are now being pursued in molecular genetics, specifically in the genes that determine the synthesis of proteins, visual pigments, and photoreceptors. I indicate directions taken to elucidate various photobiological phenomena and point out areas that still need to be investigated. Much is yet to be explored and understood in the world of light we live in. We are at the beginning of the "Age of Light."

My hope in writing this book is to awaken greater interest in the wonders that light brings to life. It is intended for students, scientists, bioengineers, and health professionals who seek greater understanding of the importance of light to living photoprocesses. Interested readers can find further enlightenment in the cited references. In the Appendix, the optics of how lenses form images is reviewed.

Pittsburgh, Pa.   J. J. W.
May 1994

Acknowledgments

The Marine Biological Laboratory and Library, as well as the Woods Hole Oceanographic Institution, Woods Hole, Massachusetts, provided just the right environment to collect my thoughts, library resources, and other courtesies during the past summers as the writing of this book progressed. I acknowledge with thanks the research support over many years from the National Institutes of Health (NIH), the National Science Foundation (NSF), the National Aeronautics and Space Administration (NASA), and the Pennsylvania Lions Eye Research Foundation.

Many of the marine organisms whose photoreceptor structure and eyes were described were collected during visiting research fellowships to the Bermuda Biological Station for Research, St. George, Bermuda; the Zoological Research Station, Naples, Italy; the University of Paris Marine Research Laboratories in Roscoff and Villefranche-Sur-Mer, France; The National Plymouth Laboratory, Plymouth, England; and The Darwin Research Station, Galapagos Islands, Ecuador. Researches were also carried out in the Department of Anatomy in the laboratory of J. Zed Young with P. Noel Dilly, University College, and Institute of Ophthalmology of London University; at The Atomic Energy Commission, Biophysics Laboratory of E. Roux, Saclay, France, and at the Institute Pasteur, Paris. I thank these institutions and scientists for their help and the many kindnesses extended to me.

I also acknowledge with thanks The Japanese Society for the Promotion of Sciences for a visiting Research Fellowship in 1988 at Tohoku University, Department of Physiology, Medical School, Sendai, Japan. This experience provided an opportunity to meet with scientists in Japan and explore with them certain aspects of photobiology and researches on vision.

As the writing of this book progressed, my debt of gratitude to Eva Keller and Ann Chang grew as they typed and retyped numerous versions of various chapters. The assistance of students Rosemary Green, Jennifer Crew, David Fergenson, and Teresa Leonardo was crucial toward organizing the numerous references, among other details of the work.

For the many services extended to me during the writing of this book, I wish express my thanks to the Department of Biological Sciences, Carnegie Mellon University, Pittsburgh, Pennsylvania. I also thank the Carnegie Mellon Photography & Graphics Department for reproducing many of the figures, and I am especially grateful to Marianne Kolson for her computer skill.

I am grateful to Professor Mary Ann Mogus for the many fruitful discussions during the writing and for reviewing various chapters, to Professors John Lindsey, John Pollock, and Dr. Dan Farkas for reviewing the book and making helpful suggestions.

The encouragement and prodding of the Oxford University Press editors, William F. Curtis and Kirk Jensen, helped immeasurably in bringing this book to publication.

I acknowledge the publishers of my previously published books for permission to reproduce certain figures and tables from: *Euglena: An Experimental Organism for Biochemical and Biophysical Studies,* Rutgers University Press, New Brunswick, NJ, 1961 (2nd revised ed., Appleton-Century-Crofts, New York, 1967); *Vision: Biochemistry and Biophysics of the Retina,* Charles Thomas and Company, Springfield, Illinois, 1966; *Photobiology,* Reinhold Publishing Company, New York, 1967; *Invertebrate Photoreceptors,* Academic Press, New York, 1971; *Photoprocesses: Photoreceptors and Evolution,* Academic Press, New York, 1975; *Liquid Crystals and Biological Structures* (with Glenn Brown), Academic Press, New York, 1979; *Light and Life Processes,* Van Nostrand Reinhold Company, Inc., New York, 1986.

Data and figures obtained from other sources are acknowledged with thanks and are referenced.

Contents

Do not take my word for it, see for yourself

—Motto of the Royal Society of London

LIGHT
AND
LIFE

CHAPTER ONE

Light and Life: An Introduction

. . . the whole of the energy which animates living beings, the whole of the energy which constitutes life, comes from the sun.
                                    —S. LEDUC, 1911, *The Mechanisms of Life*

Life as we know it on Earth is dependent on the Sun and solar radiation. Ancient civilizations throughout the world worshipped the Sun long before the advent of Western culture. The Sun was central to their religious beliefs. References to Sun worship are found in the hieroglyphics and written records of the ancient Chinese, Babylonians, Egyptians, and Native Americans. They recognized that the Sun was a source of energy, that its true color was blue—a color signifying one of the highest energies—and that blue light was intimately related to life.

A mystical sect of Hebraic origin known as the Kabbalists flourished in Persia for many centuries. Their first book of beliefs was written around A.D. 600, and by the fifteenth and sixteenth centuries they had formed an extensive belief system that linked light to creation. They believed that there was a blue light permeating the universe, forming a field of energy around all living things, and that this light was absorbed and retransmitted through the living body in a series of light emissions. A well-known biochemist and historian of science, Joseph Needham (1956), stated that it may well turn out that the "correlative thinking" of the Kabbalists had more influence on scientific minds in the dawn of modern science than has generally been credited to it. For it is truly remarkable that the intuitions and myths of these ancient societies linking the Sun's energy of blue light to life processes were not, until much later, substantiated by scientific investigation. We now know that the solar spectrum that reaches the Earth centers around 500 nm, in the blue-green of the spectrum.

Surprisingly, it was not until the nineteenth century that experimental physicists

Thomas Young (1803, 1807), J. Clerk Maxwell (1853, 1861), and Herman von Helmholtz (1867) began to lay the framework for our understanding of the physical nature of light, optics, vision, and sensory physiology.

An important discovery was made by Julius von Sachs (1864), a plant biologist studying the responses of plants to light. Von Sachs exposed plants to different colors of light and observed their phototactic responses to light. He observed that the plants bent toward the blue light. This demonstrated experimentally that plants search for light and utilize the energy of blue light for movement and for photosynthesis. Many photobiological phenomena now recognized, including *phototropism, phototaxis, photosynthesis,* and *vision,* are optimized for light around 500 nm in the blue-green. The visual spectral sensitivities of insects, birds, and rodents show response in the near ultraviolet and blue-green regions of the spectrum (Kreithen and Eisner, 1978; Jacobs et al., 1991). Experimental studies since have shown that blue light has an even more far-reaching effect on the physiological behavior of living organisms, influencing oxygen uptake, growth, pigment synthesis, and circadian rhythms. These and other effects of blue light on life's photobehavior are reviewed by Schmidt (1984) and Senger (1987).

Light influences movement, photosynthesis, vision, and behavior in living organisms. Particular wavelengths of light are extremely important for the mechanisms in the photoreceptor systems. Living organisms respond not only to blue light but also to wavelengths from the near ultraviolet into the red of the visible spectrum. Photosensors, which measure the intensity of light, the time of day, and periods of light and darkness, are found throughout plants and animals. Living organisms receive information from their environments via their receptors. In animals, the processing of this information is an essential function of the nervous system. The behavioral responses to the light–dark periods are described as *photoperiodism.* This phenomenon is observed in plants, from leaf movement to the timing of flowering, and in animals, from developmental growth, color changes in the skin, and hormonal stimulation to sexual reproduction cycles. The photoreceptors responsible for this behavior are found throughout animal bodies: in the skin, in neural cells, and even in the brain. In the human brain, photoreceptor cells found in the pituitary and pineal glands respond to daily light–dark periods and to seasonal changes which regulate hormonal secretion and sexual cycles. The skin is recognized as an important endocrine transducer of light. These receptors control an organism's overall life rhythm—its "biological clock." Other phenomena influenced by light include *photoreactivation,* the recovery of cells damaged by ultraviolet radiation, and *photodynamic action,* the photosensitization produced by the absorption of light by a molecule that becomes activated and causes photooxidation in cells.

The basis for photosensitivity in living organisms is in their photoreceptor pigment molecules, such as chlorophylls, carotenoids, flavins, and phytochromes, that are chemically structured to absorb these energies of light. Only the light absorbed by these pigment molecules is effective in promoting photochemical reactions. These pigment molecules reside within the cell, the cell membrane, and membranes of photoreceptor structures that are specialized for photoreception. For example, the chlorophyll-containing chloroplasts are the photoreceptors of plants for photosynthesis, and the rhodopsin-containing retinal rods and cones are the photoreceptors in vertebrate vision. Chloroplasts in plants and retinal photoreceptors in the eye are opto-chemo-electro devices that must have evolved early in the history of living organisms. They have been functioning effectively over the course of evolution for billions of years. The photoreceptors in these essential photoprocesses receive light energy and transduce it to chemical energy (for photosynthesis), to mechanical energy (for movement), or to electrical signals (neurosensory and visual transmission to the brain).

Certainly, our understanding of how organisms utilize light energy and convert it to chemical, mechanical, and electrical energy has been greatly advanced, but it is far from complete. Understanding these processes remains one of the great challenges in biological research. Future elucidation of these processes will require an overall grasp of an organism's energetics. Only then can we understand how an organism photosynthesizes, sees, senses, and reacts in an integrated fashion, i.e., as a whole plant or animal.

A deeper understanding of photobiological mechanisms also depends on the use of the tools of physical, chemical, and genetic molecular biology. These tools have evolved together with developments in light and electron microscopy, x-ray diffraction, and spectroscopy. With these advances in technology, scientists have begun to probe into the molecular nature of living cells. Important biochemical advances in chemistry and genetics have furthered our understanding of the molecular organization of living cells and the development of photosensory systems.

In the chapters that follow, we will identify these photoreceptor pigment molecules, their chemical structure and photochemistry, and how their photoreceptors are molecularly structured to capture the energies of light that initiate various photobiological phenomena. We will explore how living organisms detect light and their photoreceptor systems. Our discussion will eventually bring us to the eye and how images are formed. The information presented in an image depends on the physical characteristics of the optical system that forms that image. The various ways animals image their world, and the different types of optical systems they use—including simple, compound eyes, as well as refracting type eyes—will be reviewed. In the study of the different kinds of eyes, an incredible variety of optical devices for focusing and imaging will be revealed: lens systems with oddly curved surfaces, prisms, mirrors, and fiber optic light guides.

These discussions will be of more than just biological interest, for we will also examine ways in which this knowledge can be exploited to develop photoreceptor energy conversion systems, photochemical information systems, and optical imaging devices that have applications to our technology and human needs.

CHAPTER TWO

The Physical Nature of Light: The Interaction of Light with Matter and Molecules of Life

> *Light is the carrier of information—that binds us.*
> —J. BRONOWSKI, 1974, *The Ascent of Man*

THE SUN AND RADIATION

The Sun is the primary energy source for all life on Earth and supplies us with light as well as heat. The Sun produces immense amounts of energy through thermonuclear reactions, processes by which small atoms like hydrogen fuse to form larger atoms in the synthesis of helium. When this fusion occurs, the mass of the atom formed by the fused atoms is less than the sum of both atoms' initial masses. This leaves extra mass, some of which is converted to energy. The equivalence of mass and energy is well known through Einstein's relationship $E = mc^2$, in which c is the speed of light. Only about 0.7% of the Sun's mass is actually converted into energy, but because the Sun loses 5.6×10^{16} kg of matter a day, its energy output each second is equivalent to billions of the largest hydrogen bombs.

The biosphere harvests about 1% of sunlight that is incident upon the Earth. Even though the Earth receives only a small fraction of the Sun's total energy output (about 1 part in a trillion), that amount is still enormous. Of the total energy incident upon the Earth, only a small portion is retained; the remainder is radiated back (with a maximum wavelength near 11 μm in the infrared) out into space. On any given day, the amount of solar radiation reaching the Earth is roughly equal to

6

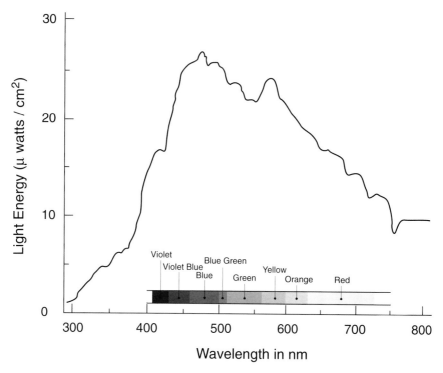

FIGURE 2.1 The solar spectrum that strikes the Earth on 82 cm² surface between 1:00 P.M. and 2:00 P.M. in Caderache, France. (Courtesy of Dr. P. Guerin de Montgareuil, Atomic Energy Commission, France.)

the sum of all energy stored since the beginning of the Earth in fossil fuels and the heat stored in the ocean waters.

In studying the radiation emitted by the Sun, we find the Sun acts very much like a perfect absorber and emitter of radiation, or a blackbody. The intensity of radiation emitted by a blackbody, for any given wavelength, depends only upon its temperature. The Sun has an average surface temperature of 5800°K, so almost all of the radiation emitted by the Sun lies in the visible range, with less radiation falling into the ultraviolet and infrared regions of the spectrum (Figure 2.1).

THE PHYSICAL NATURE OF LIGHT

> How do we answer the question, what is light? The answer is the photon, the most visible member of the family of elementary particles—that mediates the electromagnetic weak and strong interactions. The photon forms the manifestation of a symmetry principle of nature that describes the interaction of matter.
>
> —STEVEN WEINBERG, 1975, *Physics Today*

What exactly is light? This has been the subject of a major controversy over the past 300 years. Explanations before 1900 almost invariably fell into two categories: light was either a wave or a particle. Since light exhibited characteristics of both, there were always problems with either position. Let us briefly examine the historical arguments which led to our present understanding of "what is light," keeping in mind that it wasn't until this century that a seemingly simple answer was proposed: light behaves both like a wave and a particle.

Looking back, Robert Hooke (1665) and Christian Huygens (1678) described light as a wave. Isaac Newton (1704) was the first to describe light as a stream of particles. Thomas Young (1803), Augustin Jean Fresnel, and, at about the same time, Dominic Arago showed that the observed phenomena of interference patterns, polarization, and diffraction of light were of a wave nature. James Clerk Maxwell (1864) tied together the various aspects of the wave concept of light in his paper "A Dynamical Theory of the Electromagnetic Field." Maxwell's ingenious discovery is considered the basis of modern electromagnetic theory. Ludwig V. Lorenz (1867) independently evolved a similar wave theory of light, and in 1888 Heinrich Hertz (1894–1895) showed experimentally that electromagnetic waves generated by electrical circuits obey the same laws of reflection, refraction, and polarization as do light waves.

Since electromagnetic radiation, like light, may be polarized, diffracted to form interference patterns, and otherwise show qualities of being a special sort of wave, its behavior must be based on a wave-type model. Nevertheless, to explain behavior like the photoelectric effect and the fact that radiation exchanges energy with matter in bundles of discrete size, a particle description is necessary. A dual nature of light is now accepted as the proper descriptive model, although a particle acting like a wave seems strange to our macroscopic sense.

All electromagnetic radiation is composed of an electric field and a magnetic field, which, in empty space, travel with a constant velocity at the speed of light. The two fields are arranged perpendicular to each other and are constantly oscillating. It has been found that a changing electrical field will create a magnetic field, and vice versa (Figure 2.2) . The constant changes in the two fields lead to the

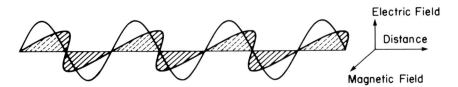

Electric Field

Distance

Magnetic Field

FIGURE 2.2 Electromagnetic waves have both an electric and a magnetic component which are perpendicular to one another. Here the magnetic field is shown in the horizontal plane, the electric field in the vertical plane, and the direction of propagation is to the left. Notice that electromagnetic waves are considered transverse waves because the component fields vibrate perpendicular to the direction of travel, which is along the axis as shown.

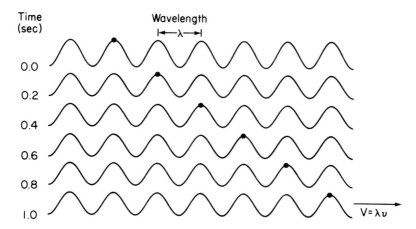

FIGURE 2.3   A transverse wave with a frequency of 5 hz (five oscillations per sec) is shown over a 1-sec period. The dot marks the travel of a single crest during this time. Since the velocity is a product of the wavelength and frequency as shown, during one second a point on the wave travels five wavelengths.

reinforcement of one field by the other, allowing them to regenerate each other if left to themselves for an indefinite length of time. This propagation is how electro-magnetic radiation in space, such as light from stars, can travel vast distances and still be visible to us.

Light, treated as a wave, is characterized on the basis of its wavelength or its frequency. The wavelength, λ, is defined in the same manner as the sine wave; one wavelength is the distance between the same point on two consecutive wave cycles. In other words, one wavelength is the distance from peak-to-peak or trough-to-trough. Another way of expressing the frequency is the number of cycles the wave goes through in one second. If at point x the wave goes from peak to trough to peak again five times in one second, then its frequency is 5 cycles per second, or 5 Hertz. If both the frequency and wavelength are known, then the velocity of the wave can be calculated (Figure 2.3). This is true because radiation is a traveling wave. If you could fix your eyes on one peak of a light wave you would see it move in space, while if you fixed your eyes on a single, immobile point in space you would see the wave oscillate from peak to trough to peak. Therefore the frequency gives the number of peaks that pass a motionless point in one second, and the wavelength gives the distance between two consecutive peaks. The distance from the first peak passing through that point, at time = 0, and the peak passing through it at time = 1 sec is simply the distance between two peaks times the frequency. In this manner we find that the distance traveled per second by a point on the radiation wave is its velocity, v. For all electromagnetic radiation this velocity is a constant, no matter how long or short the wavelength is, and is found to be 3×10^8 m/sec, which is the speed of light, c.

> *. . . modern physics, which states that events at the atomic level cannot be observed with certainty, helps resolve the paradox that particles sometimes behave like waves and waves like particles.*
> —GEORGE GAMOW, 1958, *The Principles of Uncertainty*

It was not until the beginning of the twentieth century that it was discovered that the properties of light could also be explained by the particle theory. The particle theory of light is understood through modern quantum theory. According to the theory, light is transported in wave-like bundles of energy called *photons* or *quanta*. The quantum and wave properties of radiation are not two separate qualities that together make up light; the two are intimately related. Max Planck around 1900 discovered the direct relationship between the frequency of electromagnetic radiation and the energy of its quanta. Albert Einstein (1905) extended Planck's relationship to include light. Einstein had shown—and this was a feature of Planck's derivation of the blackbody spectrum as well—that the energy of a light quantum was proportional to its frequency. That is, each photon has the energy $E = hc/\lambda$, where h is Planck's constant (6.62559×10^{-27} erg/sec), c is the velocity of light (3×10^8 m/sec), and λ is the wavelength of the light. Then the energy of a single quantum can be calculated from $E = h\nu$, since frequency is inversely proportional to wavelength and therefore $\nu = c/\lambda$. This equation shows that the higher the frequency of the radiation, the greater the energy. For example, quanta of violet light, 8×10^{14} Hertz, would be more energetic than quanta of red light, 4×10^{14} Hertz. X-rays are even more energetic, since their frequencies are higher than any of the visible, microwave, or radio frequencies.

Einstein's explanation of the photoelectric effect indicated that light, though a wave, also behaves as a particle. Louis de Broglie's (1955) theoretical studies in the early 1920s indicated that if light could be both a wave and particle, then matter could also possess wave characteristics. A general wave theory of matter that linked the subatomic world of quantum mechanics to the macroscopic world of matter was developed by Schrödinger (1928). Schrödinger's theory, for which he received a Nobel Prize in physics in 1933, explained the wave nature of matter and the probabilistic nature of the electron's behavior. This theory has enabled physicists and chemists to develop modern chemical theories of electron orbitals of the atom and the structure of molecules.

THE ABSORPTION AND EMISSION OF LIGHT

Albert Einstein (1905) proposed that all the energy of a single light quantum, or photon, can be transferred to a single electron. This one-to-one relationship between a light quantum and a particle of matter is of key importance in photochemistry. The principle that one quantum of light can bring about a direct primary photochemical change in exactly one molecule of matter is known as Einstein's Law of Photochemistry. In other words, a photon with sufficient energy can strike an electron in an atom and change the chemical properties of that atom. An excellent introduction to the quantization of the atom, spin, and molecular photobehavior can be found in the text by Eisberg and Resnick (1985).

The proper description of the absorption of light by a biological system and of light being absorbed or emitted during a reaction should contain the number of photons per second per unit wavelength. The ability of a molecule to absorb light is determined by its atomic structure, that is, by the arrangement of electrons in different orbitals about the nucleus of the atom. The electrons nearest the nucleus have relatively low energy, and those electrons in orbitals farthest away from the nucleus have relatively high energy. To move an electron from an inner orbital to an outer orbital requires energy. When photons of light strike an atom that can absorb the light, an electron in one of the orbitals may absorb the photon and gain energy sufficient to move away from the nucleus to an orbital of a higher energy level. When this happens, the atom is referred to as being in an *excited state*. In practically all cases of molecules with an even number of electrons, the photochemical behavior is describable in terms of *singlet* and *triplet* excited states. The distinction between singlet and triplet excited states of molecules that absorb light is of great importance in understanding the photochemistry of photoreceptor pigment molecules and therefore photobiology. In some atoms in the excited state, the high-energy electrons do not escape from the atom but return to their original low-energy orbitals, and the atom is said to return to the *ground state*. When the electron returns to the ground state, some of the energy is shed as (a photon or quantum of) fluorescent light.

The development of quantum mechanics in the late 1920s helped to explain how the quantum yield of energy depends upon the wavelength of the exciting light. James Franck and Edward U. Condon analyzed molecular excitation and pointed out that a molecule's transition from a ground state to an excited state takes place so rapidly that the interatomic distances in the molecule do not have time to change. The reason is that the time required for the electrons to change their shells is much shorter than the period of vibration of the atoms in the molecule.

There are four types of processes which an atom in its excited state may undergo: (1) emission of light or a radiative transition, (2) a radiationless transition between two states without chemical reaction, (3) electron excitation energy transfer, and (4) chemical reaction. For example, when a quantum of light is absorbed by a molecule, one of the electrons of the molecule is raised to some higher excited state. The excited molecule is then in an unstable condition and will try to dissipate this excess energy. Usually the electronic excitation is converted into vibrational energy, which is then passed on to its surroundings as heat. An alternative pathway is for the light excited molecule to fluoresce, that is, to emit light whose wavelength is slightly longer (lower energy) than that of the exciting radiation. An electronically excited molecule can undergo a chemical transformation and thereby can dissipate its energy. The molecule may be torn apart, as in photolysis, but this occurs only if the energy of the absorbed quantum exceeds the energy of the chemical bonds that hold the molecule together.

THE ELECTROMAGNETIC SPECTRUM OF ENERGY
AND PHOTOBIOLOGY

> *Life depends on a narrow band in the electromagnetic spectrum. This is*
> *the consequence of the way in which molecules react to radiation and*
> *must hold true not only on earth but elsewhere in the universe.*
> —GEORGE WALD, 1959, *Scientific American*

The electromagnetic spectrum of energy extends from gamma and x-rays of wavelengths less than 0.1 nm, through the ultraviolet, visible, and infrared, to radio and electric waves which are kilometers long. The solar spectrum of radiation that reaches the surface of the Earth lies between 300 and 900 nm, from the near ultraviolet through the visible and the infrared, whose maximum spectral peak is around 500 nm (Figure 2.1). The spectrum of visible light was first demonstrated by Isaac Newton (1666), who showed that by shining sunlight through a glass prism onto a screen the light separated into bands of violet, blue, green, yellow, orange, and red. These colors that we see are the frequencies, or wavelengths, of the visible spectrum from 390 to 780 nm.

Radiation in the ultraviolet, from 200 to 300 nm, is largely absorbed by ozone (O_3) in the upper atmosphere (Figure 2.4). Ozone is a good absorber of ultraviolet radiation and shields the Earth from these lethal short wavelengths. The present concentration of ozone in the upper atmosphere is sufficient to reduce the level of ultraviolet radiation at the Earth's surface by about 10^7-fold at 290 nm, but by only 200 times at 300 nm, and, fortunately for us, by negligible amounts in the visible spectrum.

Light that passes through the atmosphere to Earth also enters the sea. As in the atmosphere, light in the sea is diffused and absorbed by suspended particles,

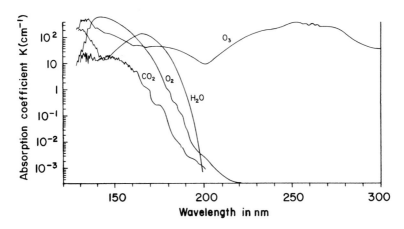

FIGURE 2.4   Ultraviolet spectral absorption of atmospheric gases. (From Wolken, 1975, and other sources.)

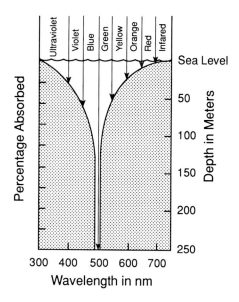

FIGURE 2.5 Solar radiation on and in water. The spectrum of energy available at each level of water is enclosed by the solid lines. Highest energy wavelengths are found in the blue-green.

sediments, and detritus, as well as by plants and animals. Below the surface of the sea, the extremes of the visible spectrum (the near ultraviolet and the red to infrared) are absorbed with increasing depth (Jerlov, 1976). At depths of 100 meters and below, the available light is narrowed to the blue-green, corresponding to the solar energy peak around 500 nm that is incident on the Earth (Figure 2.5).

Radiation in the ultraviolet is absorbed by proteins at 280 nm. Nucleic acids absorb ultraviolet light at 260 nm, which coincides with the absorption spectral peak of DNA, the genetic molecule. And, unfortunately for us, such absorption produces damaging effects on cells, greatly increasing the frequency of mutations and even cell death. In the early history of the Earth, before oxygen entered the atmosphere and formed the ozone layer that cuts out most of the ultraviolet radiation, the intensity of ultraviolet radiation must have been very high. Therefore, evolving life forms during these eras had to circumvent and repair the effects of ultraviolet damage to their DNA.

It has been shown that organisms were able to repair these damaging effects of ultraviolet radiation through a phenomenon known as *photoreactivation* with visible light. The discovery of photoreactivation by visible light was made by Kelner (1949a,b), studying bacteria, and Dulbecco (1949), studying phage. They found that after ultraviolet "killing" these organisms could be photoreactivated by blue light. The action spectrum for bacterial photoreactivation was found to have two spectral peaks, one in the near ultraviolet from about 300 to 380 nm and another in the blue from about 430 to 500 nm.

The process of a light-induced repair of DNA damaged by ultraviolet light was found to be an enzymatic photocatalysis of pyrimidine dimers (Setlow and Setlow, 1963). This unique photoenzyme combines reversibly with its substrate, the pyrimidine dimer, in damaged DNA. Upon absorption of light in the blue and near ultraviolet, this enzyme-substrate complex comes apart into the enzyme and repaired DNA. This photoreactivating enzyme probably represents the oldest of all photochemical reactions that evolved in living organisms.

Pyrimidine dimer

Photoreactivation by blue light continues to be effective, is widespread in nature, and has been observed not only in bacteria but also in algae, fungi, plants, animals, and human cells (Regan and Cook, 1969; Sutherland and Sutherland, 1975; Sutherland et al. (1980). Therefore, life forms evolving during these eras developed a way of protecting themselves against ultraviolet damage to their DNA.

Radiation in the near ultraviolet from about 300 to 400 nm and blue light from 400 to around 500 nm are of considerable interest in photobiological phenomena such as phototropism, phototaxis, and vision. For example, the spectral sensitivity for vision of most insects and some birds is from 360 to 380 nm and is around 500 to 600 nm for invertebrates and vertebrates.

In the red part of the spectrum, radiation from 600 to 700 nm is important for chlorophyll synthesis and photosynthesis. Radiation from 660 nm and into the near infrared is important for plant and animal growth, the timing of plant flowering, and sexual cycles in animals. Bacterial photosynthesis can take place even further out into the infrared, to around 900 nm. Infrared radiation beyond 900 nm is mostly absorbed by atmospheric gases, water vapor, and water that surrounds living cells.

LIGHT AND HEALTH

There are many beneficial effects of sunlight on the human body, from feelings of warmth and well-being to a restoration from states of depression. Although we have recognized the therapeutic uses of light for some time, we are just beginning to explore its numerous applications to our health and to recovery from various diseases that affect our lives.

One of the important effects of irradiation by near ultraviolet light is the conversion of provitamin D to vitamin D. The inactive precursor sterol (7-hydrocholesterol)—found in the skin of amphibians and reptiles, on the surface of feathers of birds and the hair of animals, and inside human epidermal cells—forms vitamin D naturally upon the absorption of near ultraviolet radiation to produce the active substance cholecalciferol or vitamin D. A lack of vitamin D results in the inability to deposit calcium and phosphorus in the bones. Before vitamin D therapy was known, the bone disease rickets occurred in northern climates, where there was little sunlight, but was unknown to inhabitants of southern regions. This was because the amount of sunlight is directly related to calcium metabolism and its increased absorption in bone.

About 10% to 50% of ultraviolet light is transmitted through the skin; the melanin pigments in the skin filter out much of the damaging ultraviolet radiation. While sunlight can be beneficial, constant exposure to the Sun and ultraviolet radiation produces sunburn. Chronic sunburn injury to the skin causes premature aging, and sunburn can lead to skin cancer; even a single severe burn will significantly increase the likelihood of cancer. The reason is that the action spectrum for skin cancer is around 260 nm, which is the absorption spectrum for DNA. Therefore ultraviolet absorption at 260 nm brings about structural chemical changes in the DNA molecule, which increase the frequency of cellular mutations and can lead to skin cancer. In this context, it is of interest that vitamin A and retinoids (derivatives of vitamin A) may have protective effects against the onset of certain types of tumors.

There are other damaging effects of ultraviolet and visible radiation to cells. Photosensitizing molecules are produced when a molecule absorbs light and becomes activated. These molecules can cause destructive photo-oxidation to the cell. This photobiological phenomenon is known as *photodynamic* action. Photosensitization affects many types of cells and occurs in the presence of substances that absorb the light and thereby sensitize the cell. This can be produced by naturally occurring chemical photosensitizers in the cell, by certain medications or drugs, and by pollutants in the air. But advantage can be taken of this mechanism to target photosensitive molecules in cells. As a result, light therapy is being used to treat autoimmune system diseases. The systematic changes in the immune system are an important factor in the association between ultraviolet radiation and malignancy. Near ultraviolet blue light has a dramatic effect in the treatment of jaundice in newborn infants, resulting in remarkable recovery. We are just beginning to explore and understand the medical uses of light against a variety of diseases and health-related problems (Morison, 1984; Dougherty, 1993).

REMARKS

The solar spectrum of energies that reaches the Earth lies between 300 and 900 nm, with a maximum around 500 nm (Figure 2.1). For photobiology, these wavelengths are restricted to approximately 340 to 780 nm and cluster around 500 nm, the peak of the solar spectrum. Photobiological phenomena—such as phototropism, phototaxis, photosynthesis, and vision—depend on their own range of energy to initiate these photoprocesses in living organisms. Thus, it is no accident that life on Earth has been remarkably efficient in its use of this range of energy.

Why these wavelengths of light? The reason is that the photoreceptor pigment molecules are chemically structured to absorb these energies. Hardly any photoreceptor pigment molecule absorbs wavelengths longer than 900 nm, and radiation further into the infrared is absorbed by atmospheric gases, water vapor, and water that surrounds living cells. The lower limit of the spectrum is determined by the fact that all organic molecules strongly absorb in the ultraviolet region, 240 to 280 nm, and photodamage is more likely to occur with this high energy.

Light and life became intimately tied together through their photoreceptor systems. Once the cell could efficiently trap the energies of light to perform specific functions, life and evolution continued to be possible on Earth.

With what we know about how light interacts with the molecules of matter that bring about photochemical reactions and photobiological phenomena, we will now identify these pigment molecules and how they are chemically structured for light absorption.

CHAPTER THREE

Biochromes: Pigments and Photoreception

It is a common biological conception that the occurrence of pigment in animals and plants bears a clear relation to biological effects of light.
—JACQUES LOEB, 1906, *Dynamics of Living Matter*

PIGMENTS AND PHOTORECEPTION

Photosensitivity of living organisms depends on the absorption of light by a pigment molecule or a system of pigment molecules. Living organisms synthesize pigments to capture the energy of light. What are these pigments and how are they chemically structured for photoreception? We will identify these photoreceptor pigment molecules with their biosynthesis, chemical structure, and absorption spectra. Later, we will apply this information in our discussions of how they function in the photoprocesses of photosynthesis, phototropism, phototaxis, vision, and other photobiological phenomena.

Porphyrins: synthesis and chemical structure

The basic unit of porphyrins is the pyrrole molecule. Porphyrins are structured of four pyrroles in a tetrapyrrole ring configuration. According to Calvin (1969), porphyrins were synthesized from pyrroles by a process of autocatalysis before they became incorporated into living cells. Once introduced into the living cell, porphyrins seemed to show what may be loosely termed as adaptive behavior on the molecular level. In the early history of life, protoporphyrin IX was formed by a slow and random series of reactions in primitive organisms, where it served to increase the probability that the earlier chemical steps would continue to occur. That this process, or something similar, occurred relatively early in evolutionary history is suggested by the universality of pyrrole in the organic world.

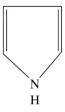

Pyrrole

Most likely, since oxidants are necessary for the synthesis of porphyrins, the porphyrins made their appearance during the evolution of organic matter, but at a somewhat later stage than adenine or flavin derivatives. Porphyrins were catalysts for chemical reactions long before they were synthesized by living organisms. They must have arisen during the transitional period when the Earth's atmosphere began to be enriched with oxygen.

The biochemical synthesis that led to the formation of porphyrins, cyto-chromes, and chlorophyll was experimentally developed by Granick (1948, 1950, 1958) and Shemin (1948, 1955, 1956). In the biosynthetic scheme, hemes and chlorophylls arise from the common precursor molecule protoporphyrin IX (Figure 3.1). These pigment molecules have a common structure of four pyrroles linked by methine bridges, $=N-\overset{\shortparallel}{C}-\overset{\shortparallel}{C}-N=$, forming a cyclic tetrapyrrole structure. The incorporation of magnesium into the nucleus of protoporphyrin IX led to the synthesis of chlorophyll. Similarly, the incorporation of iron led to the synthesis of hemes and cytochromes; the insertion of cobalt, to the formation of vitamin B_{12}. Some metal porphyrins (e.g., Mg and Zn, but not Fe) are ideal electron-transfer devices in chemical and photochemical reactions. Reactions involving such metal porphyrins are essential for sustaining life on Earth by catalyzing electron transfer between oxygen and water in both plants and animals.

Before describing in greater detail the chemical structure of chlorophylls and other photoreceptor pigments, a brief review of the structures of the respiratory pigments, hemes and cytochromes, is of interest. The heme molecule, an iron porphyrin, is synthesized from the precursor protoporphyrin IX (Figure 3.1). The heme molecule $(C_{34}H_{32}O_4N_4Fe)$ is nearly planar; the property of planarity is ascribed to the many double bonds in the molecule. The iron atom forms bonds, and is coordinated, with the four nitrogen atoms of the tetrapyrrole. Heme, with an oxygen molecule attached to the iron atom, is responsible for the red color of oxygenated blood.

The four hemes which are bound to the protein globin form *hemoglobin*. Hemoglobin is the oxygen carrier of vertebrate red blood cells and takes up oxygen

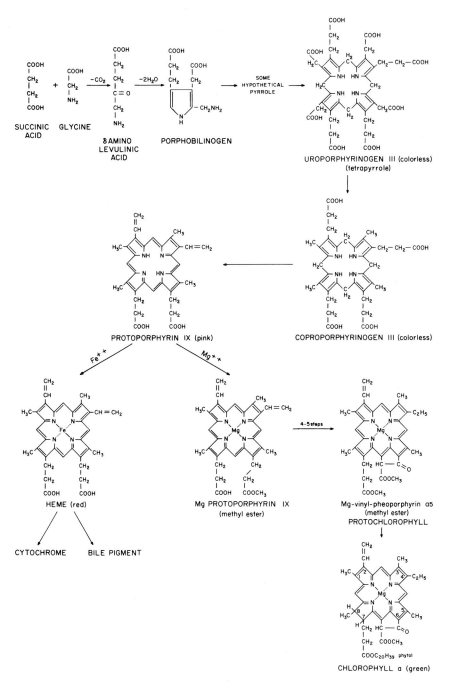

FIGURE 3.1   The biosynthesis of porphyrins, hemes, and chlorophyll (after Granick, 1950, 1958; Shemin, 1948, 1955; Calvin, 1969).

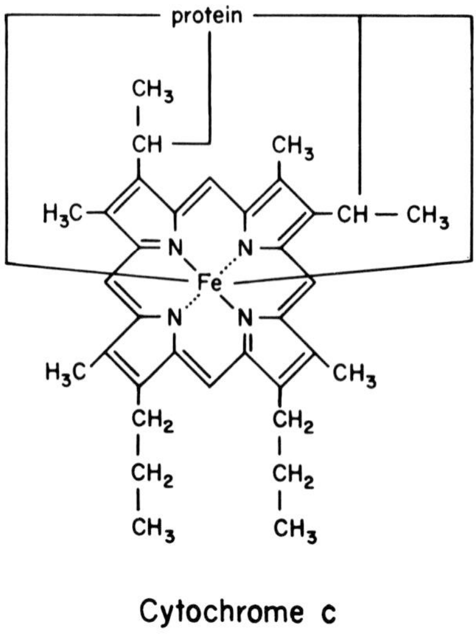

Cytochrome c

FIGURE 3.2   The chemical structure of heme-protein cytochrome *c*.

to form *oxyhemoglobin*. Although hemoglobin is more characteristic of verte-brates, it has been found in a number of invertebrates such as erythrocruorins. In many crustaceans and molluscs the blood pigment is chlorocruorin, a green pig-ment that contains copper. All the respiratory pigments have the common property of functioning as oxygen carriers.

Cytochromes are heme proteins that carry an iron atom in an attached chemical group (Figure 3.2). The red color of cytochromes comes from their prosthetic group or chromophore, which is an iron protoporphyrin IX (Figures 3.1 and 3.2). Cytochromes are found in chloroplasts and in the mitochondria of cells. They function as electron carriers during the initial reactions of the photochemical processes, and their role is that of electron transporters in the respiratory chain of oxidative phosphorylation. The various types of cytochromes are designated by the letters "a," "b," "c," and "f." These cytochromes are now distinguished on the basis of their major spectral absorption peaks, in the reduced state, for example *Euglena* cytochrome-552 (Figure 3.3). Cytochromes of the c-type have been iso-lated from bacteria, algae, green plants, and animal cells (Table 3.1). There is another class of c_3-type cytochromes that has been found in certain bacteria; this class utilizes a sulfate instead of oxygen as its electron acceptor (Pettigrew and Moore, 1987).

CHLOROPHYLL

Chlorophyll is the photoreceptor molecule for photosynthesis and is synthesized by all green plants, algae, and photosynthetic bacteria. Surprisingly, the chemical

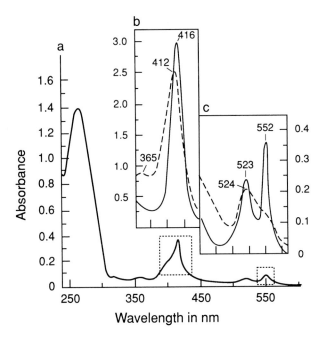

FIGURE 3.3   Cytochrome c absorption spectrum of the reduced state (a). Insert (b and c) expanded to show the major reduced absorption peaks (—) and oxidized absorption peaks (----). Isolated from light-grown *Euglena gracilis*. (From Wolken, 1967, 1975.).

structure of chlorophyll was not known until Willstater and Stoll described it in 1913. Some twenty-seven years later the structure of chlorophyll was determined by Fischer and Stern (1940), and it was not until the 1960s that chlorophyll a was synthesized in the laboratory by Woodward (1961). Chlorophyll a is a flat rigid cyclic tetrapyrrole molecule chelated with magnesium, having the empirical formula $C_{55}H_{12}O_5N_4Mg$. Its green color comes from magnesium due to the reduced double bond at the β-pyrrole position of the porphyrin molecule (Figure 3.4). The chlorophyll phytol tail is a long chain alcohol, $C_{20}H_{39}OH$, which is structurally related to the carotenoids though not conjugated (Figure 3.13).

The precursor molecule in the biosynthesis of chlorophyll is protochlorophyll (Figure 3.1). The chemical structure of protochlorophyll is closely related to that of chlorophyll. Protochlorophyll differs from chlorophyll a in that it lacks two hydrogens at positions 7 and 8 in the porphyrin ring IV of the chlorophyll molecule (Figure 3.4) and is thus an oxidation product of chlorophyll. Upon absorption of light, protochlorophyll is reduced by two hydrogens to chlorophyll a. This is observed in seeds and when etiolated plants (seedlings sprouted in darkness), which are colorless to faint green, are exposed to light and turn green. The conversion of protochlorophyll to chlorophyll is absolutely dependent on light, and the reaction occurs with a high quantum yield.

All higher plant chloroplasts contain both chlorophyll a and chlorophyll b

TABLE 3.1 Comparative Properties of Some c-Type Cytochromes

	Photosynthetic bacteria[a,b]		Algae[c]		Higher plants[b]	Animals[b]
			Euglena gracilis			
	Rhodopsirillum rubrum	*Chromatium*	Light-grown	Dark-grown	Spinach	Beef heart
Absorption maximum in nm						
Oxidized						
α	535	525	524	530	535	535
β	409	410	412	412	412	410
Reduced						
α	550	552	552	556	555	550
β	421	523	552	556	555	550
γ	416	416	416	421	417	416
Isoelectric point, pH 7	7.0	5.4	5.0	<7.0	4.7	10.0
Volts, pH 7	+0.32–0.365	+0.01–0.04	+0.35–0.40	+0.31–0.33	+0.365–0.38	+0.265
Molecular weight	16,000	97,000	11,000	13,000	110,000	13,600

[a] Data from Bartch and Kamen (1960)
[b] Kamen (1956, 1960).
[c] Wolken and Gross (1963).

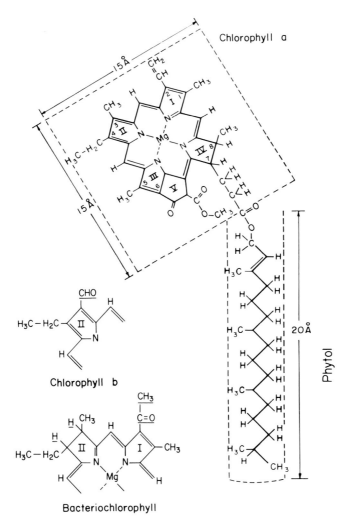

FIGURE 3.4 Chemical structures of chlorophyll *a*, chlorophyll *b*, and bacteriochlorophyll.

(Figure 3.4). Chlorophyll *a* possesses a methyl (—CH3) group at the third carbon atom, whereas a formyl (—CHO) group occupies this position in chlorophyll *b;* chlorophyll *b* is therefore an aldehyde of chlorophyll *a*. Chlorophyll *a* is present in all green plants, while chlorophyll *b* is found together with chlorophyll *a* in such plants as ferns, mosses, green algae, and euglenoids. Chlorophylls *a* and *b* differ in absorption spectra (Figure 3.5) and in their solubility. For example, chlorophyll *a* is more soluble in petroleum ether while chlorophyll *b* is more soluble in methyl alcohol. These differences in solubility make it possible to separate the two chlorophylls.

Chlorophyll a and b

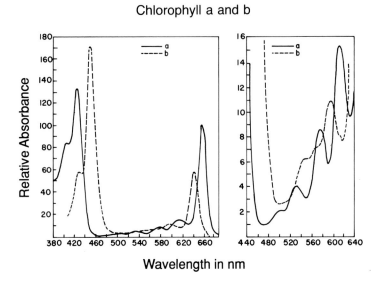

Wavelength in nm

FIGURE 3.5   Absorption spectra of chlorphylls *a* and *b* in ethyl ether; enlargement of the spectrum between 440 and 640 nm.

Other chlorophyll isomers, designated as *c, d,* and *e,* are found in diatoms, brown algae, dinoflagellates, crytomonads, and crysomonads. Chlorophyll *c* lacks a phytol group of the reduced β-pyrrole bond and is soluble in aqueous alcohol. Chlorophyll *d* is an oxidation product of chlorophyll *a* in which the vinyl group at position 2 is oxidized to a formyl group. Chlorophyll *e,* together with chlorophyll *a,* is present in small amounts in yellow-green algae. Bacteriochlorophyll is found in photosynthetic purple bacteria and differs from chlorophyll *a* in that the vinyl group at position 2 is replaced by an acetyl group which possesses two extra hydrogen atoms at positions 3 and 4 and has a second β-pyrrole bond which is reduced (Figure 3.4).

BILINS AND PHYCOBILINS

The bilin pigments are so named because they were first discovered in bile. They are metabolic degradation products of hemoglobin, hematin compounds, and chlorophyll. The name "phycobilin" indicates that they are derivable from algae. Red and green algae contain blue pigments, *phycocyanins,* and red pigments, *phycoerythrins.* Phycocyanins and phycoerythrins consist of a chromophore and a protein; because of the similarity of the chromophore, they have been termed phycobilins or biliproteins. The phycobiliprotein's chemical structure, like that of chlorophyll, is a tetrapyrrole, but an open ring porphyrin linearly arranged (Figure 3.6). The phycocyanin chemical structure is similar to phytochrome and with its

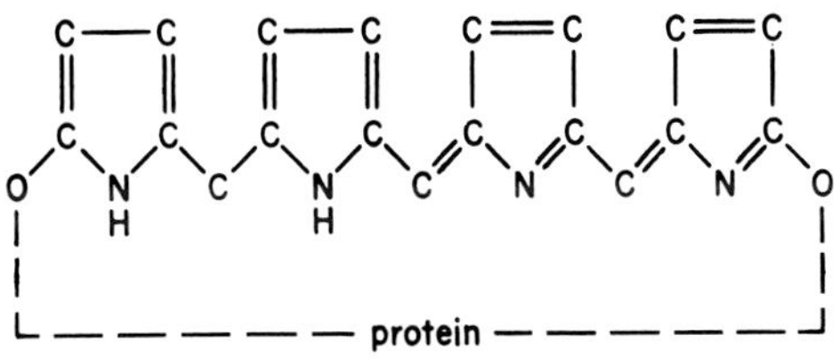

Phycobiliprotein

FIGURE 3.6   Chemical structure of four linear pyrroles of phycobiliprotein.

system of conjugated double bonds resembles the carotenoid structure (Figure 3.13). The phycobilin pigments differ from chlorophylls and carotenoids in that they are water soluble and are identified by their absorption spectral peaks.

In the green and red algae, phycobilins utilize absorbed light energy and are transferred to chlorophyll *a* in the process of photosynthesis with an efficiency equivalent to or greater than that of chlorophyll alone.

PHYTOCHROME

Phytochrome is a chromoprotein, which, like the phycocyanins, is a linear tetra-pyrrole. The chemical structure of phytochrome and its spectral absorption are shown in Figures 3.7 and 3.8, respectively. The pigment phytochrome was discovered in plants, and its photochemical function in plant behavior has an interesting history.

It has been observed since ancient times that the leaves of certain plants fold at night and open in the morning. Plants will orient themselves to varying intensities of light during the day and in darkness. In effect, they measure the amount of light and "clock" the time of day; they are *photoperiodic*. Photoperiodism is exhibited by plants in their response to variations in the intensity of light and to different wavelengths of light. These factors control germination, growth, and flowering. For example, continuous red light around 660 nm was found to be effective in altering a plant's response, that is, in preventing flower formation. Furthermore, a flash of red light during the subjective night for the plant could also alter flowering. However, it was also found that if a flash of red light was followed immediately by a short interval of far-red light (730 nm), the effect of the flash of red light was negated. The plant then acted as if its nighttime had never been interrupted; that is, it flowered. The sensor pigment responsible for this red and far-red effect is *phytochrome* (Hendricks, 1968). Phytochrome has two light reactions with distinct states, one with a maximum absorption in the *red* near 660 nm P_r and another with a maximum absorption in the *far-red* near 730 nm P_{fr} (Figure 3.8). Phytochrome is a switching device in which the two forms of the molecule P_r and P_{fr} are interconvertible. Instead of facilitating long-distance transport, it serves to

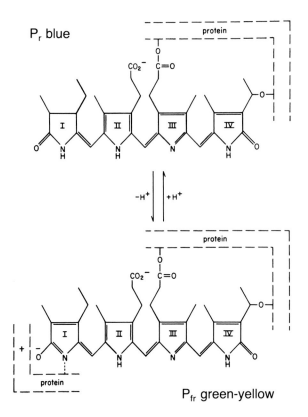

FIGURE 3.7   Chemical structure of phytochrome. Note the similarity in structure to phy-cobiliprotein (Figure 3.6).

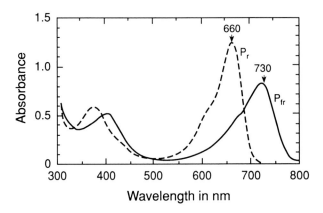

FIGURE 3.8   Phytochrome absorption spectrum of the red P_r form (----) and the far-red P_{fr} form (—), isolated from oats.

create an adjustable absorption band at a relatively low quantum energy in the red or far-red of the spectrum as indicated:

$$P_r \underset{730 \text{ nm}}{\overset{660 \text{ nm}}{\rightleftharpoons}} P_{fr} \xrightarrow{\text{darkness}} P_r$$

The phytochrome transformation from the red absorbing form to the far-red absorbing form involves a change in the membrane protein conformation, that is, a change in molecular shape that alters the receptor membrane properties to which it is bound. Hendricks and Siegleman (1967) proposed a mechanism for the photo-transformation of phytochrome based on a *cis* to *trans* isomerization, a flip-flop between two isomeric states similar to that of retinal in rhodopsin, the photosystem for visual excitation.

There is no experimental evidence to date that phytochrome is present in animals, though researchers have been looking for a similarly structured molecule in animals that functions in photoperiodic behavior.

FLAVINS AND FLAVOPROTEINS

Flavins and flavoproteins, of which riboflavin (vitamin B_2) is an example, are yellow photosensitive pigments. Riboflavin is synthesized by a number of micro-organisms and by most higher plants and is found in nearly all animal tissues. As one of the B vitamins, it is a factor used in cellular respiration.

A riboflavin molecule consists of D-ribitol attached to a substituted isoalloxazine ring, and the chemical structure is benzisoalloxazine-6,7-dimethyl-9 D-ribitol (Figure 3.9). Riboflavin in solution is yellow and can change reversibly

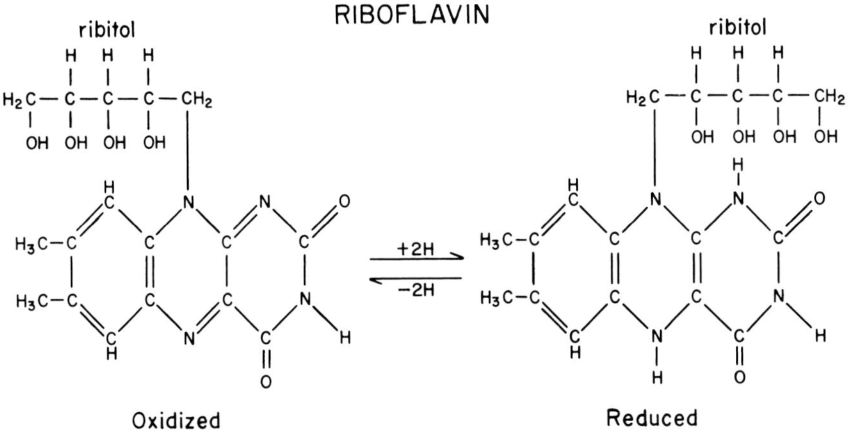

FIGURE 3.9   Chemical structures of riboflavin (vitamin B_2).

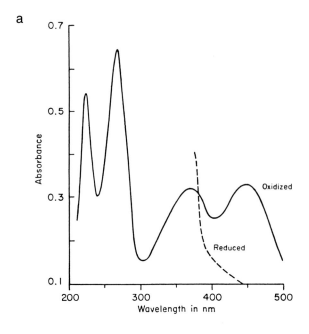

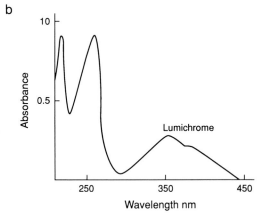

FIGURE 3.10   (a) Riboflavin absorption spectra oxidized (—) and reduced by dithionite (----) states. (b) Lumichrome absorption spectrum.

from the yellow-colored, oxidized form to the colorless, reduced form. The spectral absorption peaks in the oxidized state are around 221 to 227, 265 to 270, 365 to 370, and 445 to 460 nm (Figure 3.10a). All biologically relevant flavin reactions are then oxidation–reduction reactions. Upon ultraviolet excitation, riboflavin fluoresces blue-green, whose emission spectrum is around 520 to 560 nm. Riboflavin is light sensitive in neutral or in acidic solution and is photosensitized

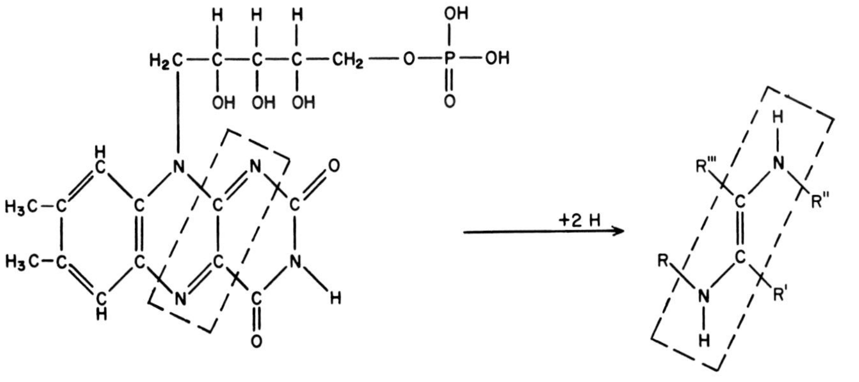

FMN   **FMNH₂**

FIGURE 3.11   Chemical structures of flavin mononucleotide (FMN) and its reduced form, FMNH$_2$.

to *lumichrome,* whose absorption peaks are at 223, 260, and around 360 nm (Figure 3.10b).

Riboflavin mononucleotide (FMN) is riboflavin-5′-phosphate (Figure 3.11). Warburg and Christian (1938 a,b,c) discovered that D-amino oxidase, an enzyme catalyzing the oxidation of D-amino acids, was a flavoprotein containing a prosthetic group distinct from FMN; this coenzyme is flavin adenine dinucleotide (FAD).

Flavins and flavoproteins absorb strongly in the blue. Flavins have been identified as a photoreceptor for chloroplast movement. They also participate in the chloroplast electron-transport system during photosynthesis. Flavins are photoreceptor molecules associated with photoprocesses of phototactic behavior, phototropism, and phototaxis of fungi, algae, and protozoa. Flavins are found in the retina of mammalian eyes, in the pigment epithelium of fish, and between the pigment epithelium and choroid of frog, rabbit, rat, and bovine retinas. They are not the photoreceptor molecules for animal vision. They do, though, function in the biochemistry of photoreception and in visual processes.

PTERINES

The name "pterine" is derived from "lepidopterine," since it was first found in the wings of butterflies by Sir Frederic Gowland Hopkins (1889) and accounts for the yellow coloration in the wings of many types of insects. These wing patterns are disguises to mimic their environment so as to attract prey or ward off predators—a kind of adaptive selection by insects observed in butterflies, moths, and wasps. Pterines are also found in association with the photoreceptor system for phototactic behavior of fungi, algae, and protozoa (Brodhun and Hader, 1990). Pterines, or pteridine derivatives, are the red and yellow pigments in insect eyes (Forrest and Mitchell, 1954). Although they participate in some metabolic processes in the eye, they do not appear to be a primary molecule for visual photoreception.

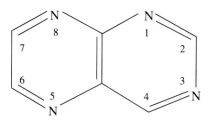

Pteridine

Comparison between the two alloxazine ring structures of pteridine and the three rings of flavins shows a close chemical structural relationship (Figure 3.9). Certain pteridine complexes with protein may serve as coenzymes analogous to that of flavoproteins.

Pterines are stable *in vivo* but photosensitive *in vitro*. Pterine pigments are usually chemically represented as the yellow xanthopterin (2-amino-4, 6-dihydroxypyrimido-pyrazine) ring, and, upon ultraviolet excitation, they fluoresce blue.

MELANINS

Melanins are a ubiquitous class of biological pigments. They range in color from yellow to reddish brown and from brown to black and are a principle pigment of vertebrates. In humans, the degree of pigmentation varies with the extent of exposure to solar radiation and with the aging process. The nature of these pigments is determined by the genetic make-up of the organism and the biological function of the pigment. These functions include camouflage from predators and sexual recognition and attraction within species.

Melanin pigments are found in the hair, skin, and eyes. The main function of melanin in the skin is to act as a screening pigment by protecting the skin cells, and, in the eye, it protects the retina from ultraviolet radiation damage.

Melanin is a biochrome of high molecular weight and can be produced *in vitro* by the oxidation of the amino acid tyrosine with the enzyme tyrosinase, a copper-containing protein. The first step in this reaction is the production of DOPA (3,4-dihydroxyphenylalanine), which is oxidized enzymatically to DOPA-quinone (Figure 3.12); afterwards a complicated series of further oxidations and polymerizations occurs, leading to the formation of tyrosine-melanin. This is observed by noting how a cut apple, potato, or banana turns brown-black on exposure to air. The exact chemical structure of natural melanin that is synthesized by the chroma-

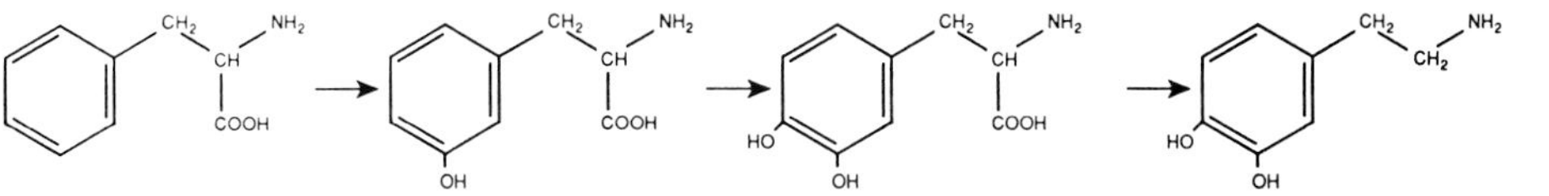

Phenylalanine   Tyrosine   3, 4-dihydroxyphenylalanine   3, 4-dihydroxy-phenyl-methylamine
                                                        (DOPA)   (DOPamine)

FIGURE 3.12   The chemical pathways in synthesis from phenylalanine to tyrosine to DOPA leading to melanin.

tophores, or the melanophores and the melanocytes of cells, is complicated by the fact that melanin is always bound to a protein. However, all experimental evidence at present indicates that tyrosine is the precursor molecule of melanin. The biosynthesis follows the pathway of tyrosine → DOPA → melanin.

CAROTENOIDS

Carotenoids are yellow, orange, and red pigments that are widely distributed in living organisms. They are synthesized by bacteria, algae, fungi, and plants and serve a wide variety of functions. In green plants, carotenoids are found together with chlorophyll. They are especially important in photosynthetic systems where they have the dual function of light harvesting and photoreception (Goodwin and Britton 1988). For example, all plants and animals that exhibit phototropism, phototaxis, and vision have been shown to depend upon carotenoid molecules or their derivatives for photoreception. The evidence for this is that their action spectrum (spectral response to behavior) resembles the absorption spectrum of carotenoids.

The structure of carotenoids is a system of single and double bonds and consists of forty carbon atoms composed entirely of carbon and hydrogen hooked together in a long polyene chain (Figure 3.13). The system of conjugated double bonds enables the carotenoids to absorb light in the visible. The carotenoids are named for their most familiar substance, carotene, and are divided into two main groups: the *carotenes* (pure hydrocarbons), the most abundant of which is all-*trans*-β-carotene, $C_{40}H_{56}$, and the *xanthophylls*, $C_{40}H_{56}OH_2$ (oxygen containing derivatives). One of the common xanthophylls is lutein, $C_{40}H_{56}(OH)_2$, or luteol. The oxygen atoms can be in hydroxyl, carboxyl, or methoxyl groups. Based on their chemical structure, carotenoids can be considered to be built from isoprene (2-methyl-1, 3-butadiene) units. In its linear arrangement, a carotenoid molecule consists of four radicals of isoprene residues. The isoprene units are linked so that the two methyl groups nearest the center of the molecule are in positions 1 and 6, while all other lateral methyl groups are in positions 1 and 5 (Figure 3.13).

The carotenoid molecule consists of a chromophoric system of alternating single and double interatomic linkages between the carbon atoms, a polyene chain

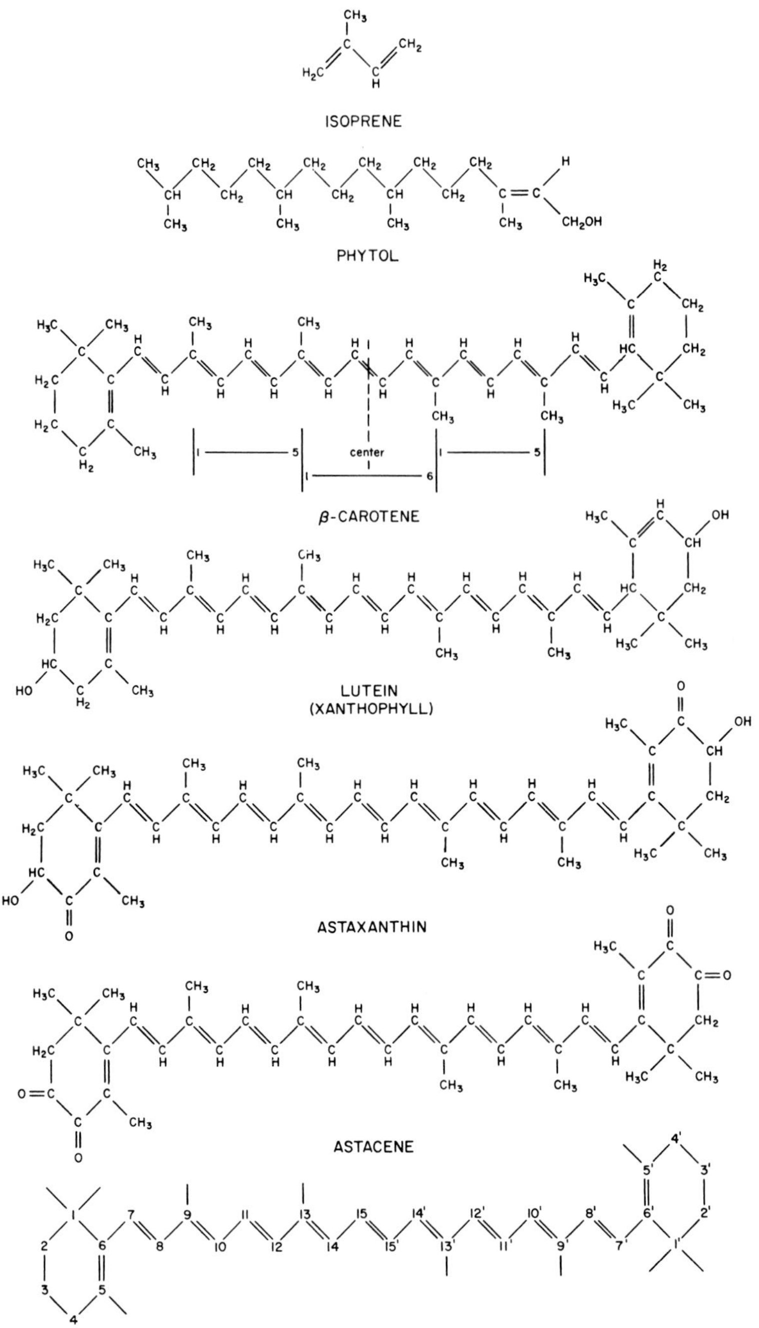

FIGURE 3.13   Chemical structures of carotenoids. Carbon numbered according to Karrer (Karrer and Jucker, 1950).

of conjugated double bonds. The large number of these conjugated double bonds offers the possibility of either *cis-* or *trans-*geometric configurations (Zechmeister, 1962). It is estimated that there are about twenty possible geometric isomers of β-carotene, of which six *cis* isomers have been discovered in nature. The spectral characteristics, and therefore the color of the carotenoid, are largely determined by the number of conjugated double bonds in the molecule.

Carotenoids are generally associated with the 20-carbon aliphatic alcohol, *phytol* (Figure 3.13), which is the colorless moiety of the ester-comprising chlorophyll. The striking resemblance between the carotenoid skeleton and phytol holds also for the details of spatial configuration. In the biosynthesis of carotenoids, mevalonic acid is a common precursor, as indicated in Figure 3.14. Geranylgeranyl pyrophosphate (GGPP) is synthesized from mevalonic acid by the normal terpenoid pathway and has now been established as the immediate precursor to phytoene, the first C_{40}-carotene (Liaaen-Jenson and Andrews, 1972). Neurosporene is the common precursor for α-, β-, γ-, and δ-carotene. The oxygenated carotenes, xanthophylls, are derivable from β-carotene.

Animals cannot synthesize C_{40} carotenoids and need to obtain β-carotene by ingesting plants. The ingested β-carotene is metabolized by animals to a degraded derivative of the C_{20}-carotenoid molecule, vitamin A (retinol) (Figure 3.15). For example, animals convert β-carotene ($C_{40}H_{56}$) to vitamin A ($C_{19}H_{27}CH_2OH$), an alcohol whose terminal aldehyde (CHO) retinal is the chromophore of the visual pigment rhodopsin. The chemical structure and photochemistry of rhodopsins in eyes of animals are discussed in Chapters 7 and 9.

ABSORPTION AND ACTION SPECTRA

To identify the pigment or pigments associated with a photoprocess requires solvent extraction or chemical purification of the pigment from the organism's photoreceptors. Further purification by physical-chemical methods or chromatographic may be required to isolate the pigment molecule. Once isolated, the pigment is then dissolved in pure organic solvents and identified spectroscopically by its absorption spectral peaks. The absorption spectral data of the various pigments are shown in Absorbance versus Wavelength. In measuring the absorbance, $A = \log_{10} I_0/I$, where I_0 is the intensity of the entering wavelength of light and I is the intensity of the wavelength of light transmitted through the pigment solution. Absorption spectra of purified pigments are of great informational value in establishing identity and chemical structure.

In the absence of the ability to isolate and identify the photoreceptor pigment molecules, clues to identity can be obtained from the *action spectrum*. The action spectrum for a photoprocess is determined by measuring the behavioral response of an organism to a light stimulus of various wavelengths and light intensities. The spectrum obtained for the behavior should correspond to the absorption spectrum of the photoreceptor pigment molecule responsible for this behavior. The action

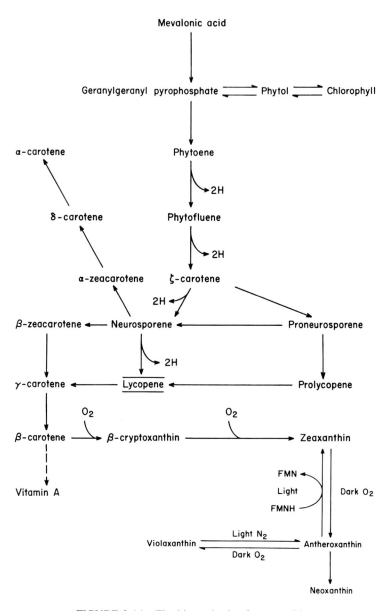

Biosynthesis of Carotenoids

FIGURE 3.14   The biosynthesis of carotenoids.

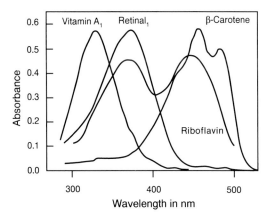

FIGURE 3.15   Comparison of absorption spectra of vitamin A, retinal, β-carotene, and riboflavin.

spectrum can then be compared to the absorption spectrum of the extracted pig-ments or compared with the spectral data available for known pigments. An example can be seen in the action spectrum for chlorophyll synthesis compared to the absorption spectrum of chlorophyll and its precursor, protochlorophyll (Figure 3.16). Another example can be seen when the spectral sensitivity of the human eye is compared to the absorption spectrum of the visual pigment, rhodopsin (Figure 3.17).

   Therefore, the action spectrum is an extremely powerful tool for identifying a

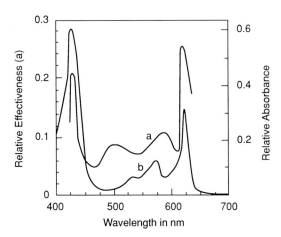

FIGURE 3.16   Action spectrum. Relative effectiveness for chlorophyll synthesis spectrum (a) compared to absorption spectrum (b) for protochlorophyll.

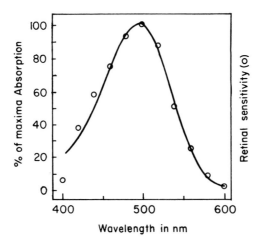

FIGURE 3.17   The action spectrum of the spectral sensitivity of the human eye compared to the absorption spectrum of the visual pigment rhodopsin. (From Crescitelli and Dartnall, 1953.)

photochemical process with a specific photoreceptor pigment molecule in the study of photobiological phenomena.

CONCLUDING REMARKS

The pigment molecules that function for photoreception in photobiology are surprisingly few in number. They are principally chlorophyll, phytochrome, flavins, retinal, and the photoreactivating enzyme. Why are these photoreceptor pigment molecules relatively limited when there are unlimited numbers of molecules that function in cellular biochemistry without light, and why were these molecules selected out in the process of evolution to absorb the visible bands of energy?

An answer can be found in their chemical structures. It was noted that, for the most part, these molecules are derived in their biosynthesis from common precursor molecules. These pigment molecules are structured of carbon to carbon bonds, linked by alternating single and double bonds (—C=C—C=C—, etc.), *conjugated polyene chains*. A common chemical structure is found in carotenoids, in chlorophylls, and in the linear tetrapyrroles phycobilins and phytochromes. The absorption spectra of these pigment molecules absorb in the visible spectrum, and many photobiological phenomena cluster around the solar energy peak 500 nm, in the blue-green.

A quantum of visible light around 500 nm represents a very large amount of chemical energy, about 2eV or 46 kcal/mole. This energy must be utilized by the photoreceptor pigment molecules; once degraded these molecules must be reconstituted and not permitted to become randomized. Living organisms accomplish this via enzymes that catalyze the conversion of $2 \, O_2 + 2 \, H^+ \rightarrow O_2 + H_2O$. Therefore, the photoreceptor pigment molecules of aerobic cells must be structured to minimize a quantum of light energy. To do so, accessory pigment molecules that participate in these photoprocesses can absorb and dissipate this energy, for example carotenes such as β-carotene (Krinsky, 1971; Delbrück, 1976).

With the information we have on photoreceptor pigments, their chemical structures, and their absorption spectra, relationships can be deduced between these pigment molecules which absorb and transduce radiant energy to the photoprocesses of living organisms.

The pigment molecules for photoreception reside in the cell, in the cell membrane, and in the membranes of photoreceptors. Therefore, it is of interest to turn to the molecular structure of the cell membrane before discussing the photoreceptor structures in which these pigment molecules reside.

CHAPTER FOUR

The Cell Membrane: Molecular Structure

> *The physiologist finds life to be dependent for its manifestations on particular molecular arrangements.*
>
> —THOMAS HUXLEY, 1866

Cells are the basic units of life. Cells are enclosed by a cell membrane that encapsulates all the intracellular components, *organelles,* of the cell. The cell membrane provides the means for cells to separate their external environment from their internal environment. The cell membrane has selective properties and allows for the differential diffusion of ions and the exchange of gases. The cell membrane was at one time envisaged as only a passive barrier for diffusion and permeability, but it is now known to play an active role in chemical transport, energy transduction, and information transfer to and from the cell.

Since the cell membrane is so important to the integrity of the cell and to biochemical processes necessary for life, it will be informative to first review the molecular structure of the cell membrane, then to see how the photoreceptor pigment molecules are molecularly associated with the cell membrane for photoreception and how they function in photoprocesses of living cells.

The cell membrane structure depends on the physical-chemical properties of lipids (Table 4.1). The most abundant of the naturally occurring lipids in the cell are lecithin, phosphatidylcloline, cephalin, and phosphatidylethanolamine. Their basic chemical structure is illustrated in Figure 4.1. For most membranes, lipids (phospholipids and sterols) are found in concentrations greater than 30%. The lipids have the unique property of forming mono- and bimolecular layers when dispersed in water. This is due to the presence of hydrophilic (water soluble)

TABLE 4.1   Lipids in Cellular Membranes[a,b]

	Myelin	Erythrocyte	Mitochondria	Microsome
Cholesterol	25	25	5	6
Phosphatidylethanolamine	14	20	28	17
Phosphatidylserine	7	11	0	0
Phosphatidylcholine	11	23	48	64
Phosphatidylinositol	0	2	8	11
Sphingomyelin	6	18	0	0
Cerebroside	21	0	0	0

[a] Data taken in part from Korn (1964).
[b] Values given in percent.

groups at one end of the molecule and hydrophobic (fat soluble) groups at the other end of the molecule.

Plants and animals are composed of cells that are assembled into highly specialized organs in an integrated system that functions like a "machine." The cell possesses specialized "organelles" enclosed by membranes to carry out metabolic chemical processes for growth and replication. Cellular organisms also have sensors (receptors that detect and measure the effects of light, heat, and pressure) and photoreceptors that respond to chemical and to these physical forces in their environment.

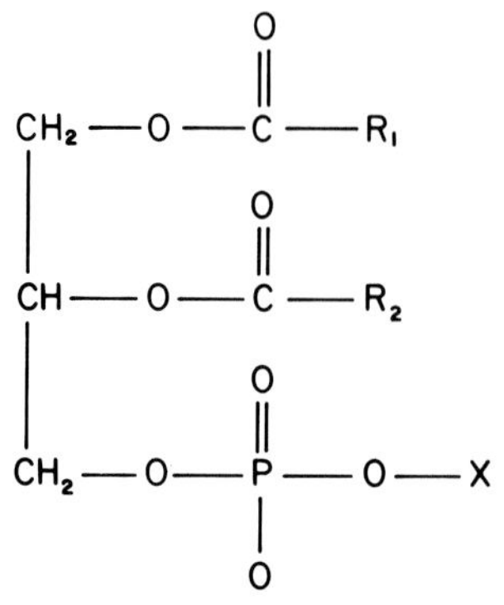

Phospholipids	X
Phosphatidyl choline (Lecithin)	$-CH_2-CH_2-N^+(CH_3)_3$
Phosphatidyl ethanolamine (Cephalin)	$-CH_2-CH_2-NH_3^+$
Phosphatidyl serine	$-CH_2-CH(NH_3^+)-COO^-$
Phosphatidyl inositol	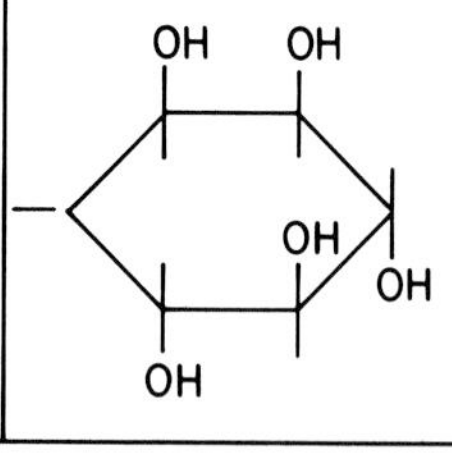

FIGURE 4.1 Basic chemical structure of common phospholipids. R_1 and R_2 are alkyl radicals with chain lengths of 16–20 carbons.

In water the charged phosphates face outward, and if the medium is nonpolar, they face inward (Figure 4.2a–d). In other words, the polar groups orient toward water or other polar molecules, and the nonpolar groups orient away from the polar environment. Phospholipids swell in water and form spherical bodies composed of concentric layers (*lamellae*) with water trapped between them. If the spheres are surrounded by a single phospholipid bilayer, they are referred to as "liposomes."

For example, lecithin dispersed in water will form concentric, bilayered lamel-

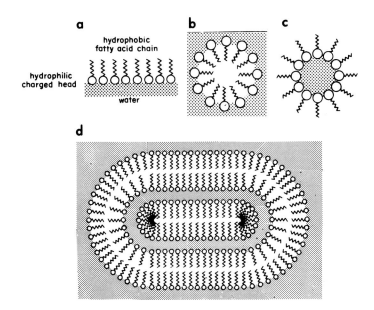

FIGURE 4.2   (a) A monolayer of phospholipid molecules in water. The phospholipids are symbolized by a circle representing the charges' hydrophilic end, and the zig-zag line represents the hydrophobic fatty acid chain. (b) If the liquid is polar, like water, the charged phosphates face outward. (c) If it is nonpolar, like benzene, they face inward. It can also exist as a combination of (b) and (c) as indicated in (d), typical of lipid bilayer membranes.

lae, which are observed as myelin structures in cells. Lecithin molecules in physiological saline will self-assemble into ordered, replicating bilayer structures as do all cellular membranes (Figure 4.3). Therefore, the molecular packing of the lipids dictates the skeletal structure of the membrane.

Cell membranes studied by electron microscopy and x-ray diffraction indicate that the cell membrane structure is a bilayer, 100 Å in thickness, each layer of the bilayer being about 50 Å in thickness (Figure 4.4). Proteins and enzymes are molecularly associated with the lipid bilayers. The proteins on or in the lipid bilayers allow the membrane to carry out transmembrane reactions, that is, chemical transport, energy transfer, and signal transduction. Therefore, every type of membrane has a unique set of proteins and enzymes to account for its function. Membrane proteins, except for a few, are not well characterized, nor are their molecular structures precisely known. In cellular membranes there are differences between the total lipids and the total proteins that constitute the membrane, but many membranes approach a 50:50 ratio (Table 4.2). To know how the protein molecules are structured within the lipid bilayers and how they are oriented in relation to the molecular structure of membranes is necessary to an understanding of how they function.

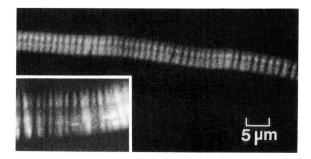

FIGURE 4.3   Membrane formed from lecithin in physiological saline, photographed by phase contrast microscopy using polarized light and a quarter wavelength filter. (From Wolken, 1984.)

The early work of Danielli and Davson (1935) and that of Robertson (1959) visualized the membrane as a lipid bilayer in which a functional protein formed continuous layers on the surfaces of the lipid bilayers and was referred to as the *unit membrane*. Singer (1971) though pointed out that the Danielli and Davson and Robertson membrane models had conceptual difficulties and could not explain many experimental results. As a result, Singer proposed a model in which the proteins penetrated deeply into or through the lipid bilayer to form a "mosaic." Such proteins are amphophilic and held in the bilayer through hydrophobic interactions. Singer and Nicholson (1972) extended this model to the "fluid mosaic" model, depicting the lipid phase of the membrane as a two-dimensional liquid in which both protein and lipid molecules could diffuse freely. This model of the membrane possesses many appealing properties. The protein can float in the liquid since lipids and proteins are mobile when in a fluid (or melted) state. Also, both fluid and solid regions may be present in the same membrane, and Oldfield (1973) indicated that membranes contain some of their lipids in a crystalline state. These phospholipid mono- and bilayer structures are liquid crystalline systems (Brown and Wolken, 1979; Chapman, 1979). That is, these lipids in the membrane undergo phase transitions from a fluid to a crystalline state.

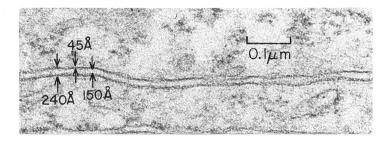

FIGURE 4.4   Cell membrane structure electron micrograph.

TABLE 4.2 The Percent of Lipids to Proteins in Various Cellular Membranes

Membrane	Lipid %	Protein %
Chloroplast	50	50
Chloroplast (Thylakoids)	30	70
Mitochondrion (outer membrane)	48	52
Retinal rods (outer segments)		
Bovine	38–49	51–62
Frog	41	59

Data taken in part from Datta, 1987; Korn, 1964, 1966; Wolken, 1975, 1986.

Other modifications of the membrane protein models have been proposed. Vanderkooi and Green (1971) suggested that some membrane proteins might be bimodal and not electrostatic. That is, their protein molecules would possess both polar and nonpolar groups like the phospholipids. Proteins in a globular configuration would fit directly into a lipid bilayer, their hydrocarbon chains and their polar groups bonded to lipid heads. Freeze-fracture electron microscopy and spectroscopy have directly demonstrated the presence of proteins in the interior of nearly all membranes and have confirmed the diffusion of both lipids and proteins in the plane of the membrane (Tanford, 1980).

Membranes are not static structures; they exist in a dynamic state, and their molecules have mobility. Labeling experiments show that although the membrane molecules do not exchange rapidly from one side of the bilayer to the other most other motions are possible. It appears that the membrane has a fluid lipid matrix in which embedded molecules can move rather freely. One can say that a membrane containing phospholipids with little unsaturation is less fluid than one that is greatly unsaturated. The control of the fluidity of the components of cell membranes may be related to the diffusive characteristics of molecules and ions passing in and out of the membrane. The state of the phospholipid in a membrane, in a gel, or in a liquid crystalline state can be expected to have a marked effect on the function of the membrane. Thus, small molecules will be able to move relatively easily through a membrane in which the phospholipids are in a liquid crystalline state.

The dynamic nature of the membrane is observed, for example, in the fast molecular motion about the C—C bonds in the lipid hydrocarbon. The lipids can exhibit translational motion and can undergo fast lateral diffusion within the plane of the membrane. The degree of freedom possible for the phospholipid chains in the membrane is illustrated in Figure 4.5. Although lateral diffusion of lipids is known to occur rapidly, different lipid orientation is possible in different regions of a membrane (Bergelson and Barsukov, 1977). Proteins show comparable motions but are slower. The membrane also has static features, for not all membranes

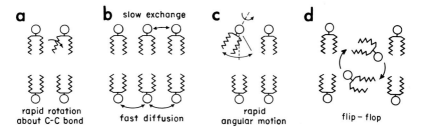

FIGURE 4.5   Schematic showing mobility of the phospholipid hydrocarbon chain in the cell membrane. (a) Rapid rotation about the C—C bond in the hydrocarbon. (b) Lateral diffusion in the plane of the membrane. (c) Angular motion of the phospholipid molecule. (d) Flip-flop of the phospholipid molecules across the bilayer. (From Brown and Wolken, 1979.)

exhibit highly fluid properties; the degree of fluidity will depend on the lipid composition. In the boundary layers where the lipid is dissolved in water, the structure is rather immobile. The proteins can also be mobile, but their motional properties in the membrane will depend on the lipids which surround a specific protein and the degree of interaction between the lipid and other proteins in the membrane, as well as with other proteins in the cytoplasm and cytoskeleton.

There are many ways in which the proteins and receptor molecules are associated with the lipid bilayer in cellular membranes. They can be associated on the surfaces of the lipid bilayers, through the lipid bilayers, and between the lipid bilayers (Bretscher, 1973; Capaldi et al., 1973; Eisenburg and McLaughlin, 1976; Meyers and Burger, 1977). These are schematized in Figure 4.5 for the various cell membranes which fit with the experimental data at our present level of understanding. In Figure 4.6a, the proteins are on the surfaces of the lipid layers; proteins can also extend into the lipid layer (Figure 4.6b) and are capable of moving through the lipid bilayer by rotational and lateral diffusion, (Figure 4.6c), or a protein molecule can extend through the entire width of the lipid bilayer (Figure 4.6d). Beside proteins residing on the surfaces of the lipid layers, another protein can lie between the lipid layers (Figure 4.6e). This model is further expanded to show that another protein can wrap around the surface proteins as in an α-helix (Figure 4.6f). These models indicate that a number of possibilities exist for proteins.

Membranes hold a number of advantages which can be exploited for photoreception (Wolken, 1975, 1986). They provide an interface between the external and internal environments. Their structure is associated with a unique property of lipids, that is, their ability to form mono- and bimolecular layers which can serve to separate one part of a reaction from another. When the membranes are folded into lamellae, their volume is minimized but their surface area is maximized for all the photoreceptor pigment molecules. Membranes not only provide a large surface area for photoreceptor pigment molecules, they bring the molecules close together

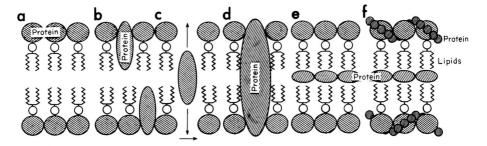

FIGURE 4.6   Membrane models. (a) The lipid bilayer with associated proteins. (b) The lipid bilayer in which proteins are not only on the surfaces but also between the bilayer. (c) The protein with rotational reaction as well as lateral diffusion. (d) A membrane in which a protein extends through the lipid bilayer. (e) The lipid bilayer in which proteins are not only on the surface but also between the lipid bilayer. (f) A membrane similar to (e) but with another protein wrapped in an α-helix around the surface protein.

for the orientation and interaction of substrate molecules and provide sites for enzymatic reactions. Since the membranes are closely packed as in a crystal, they bring the receptor molecules within molecular distances for interaction.

What is apparent, then, is that membranes are organized into a two-dimensional crystalline lattice and that such a structure is an efficient mechanism for energy capture, amplification, and regulation essential to the process of photoexcitation. Membranes provide the basic molecular structure for the photoreceptor pigment molecules and, hence, photoreception. With this in mind, we will first review the photoreceptor, the chloroplast molecular structure to photosynthesis. In later chapters, the photoreceptor structures that initiate photosensory behavior and vision will be described.

CHAPTER FIVE

Specialized Cellular Membranes for Photoreception: The Chloroplast in Photosynthesis

The molecules out of which living material is made contain large stores of internal energy. . . . And from what source do the molecules of living creatures here on Earth get their internal supplies of energy? . . . Plants get it from sunlight, and animals get it from plants, or from other animals. . . . So in the last analysis the energy always comes from the Sun.
—FRED HOYLE, 1967, *The Black Cloud*

Steven Hale presented his studies on cucumbers to the Royal Society of London in 1727. Jonathan Swift seized upon Hale's report, embellishing it to produce what he perceived as a ludicrous notion for one of the satirical experiments described in his *Gulliver's Travels*, 5th chapter of Part II. There, one of the resident scientists in the Academy of Lagado is busily engaged in an eight-year project to extract "sunbeams" out of cucumbers, put them into hermetically sealed vials, and let them out to warm the air. The irony, of course, is that in burning fossil fuels and in fermentation processes, we have been unknowingly "extracting sunbeams" from plants for ages.

Photosynthesis is crucial to life. It is a process that converts solar energy directly to chemical energy (Figure 5.1). Photosynthesis is ultimately the source of food for all animal life and a vital process for sustaining life on Earth. Photo-

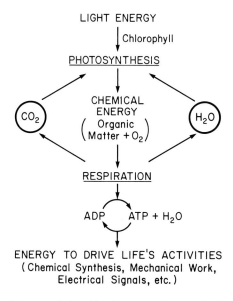

FIGURE 5.1   Energy relationships between photosynthesis and respiration.

synthesis evolved in bacteria, algae, and plants that were capable of converting the light energy to chemical energy via chlorophyll by synthesizing organic compounds according to the fundamental equation:

$$CO_2 + H_2O \xrightarrow[\text{chlorophyll}]{\text{light}} \underset{\substack{\text{organic} \\ \text{compounds}}}{(CH_2O)n} + O_2 \uparrow$$

Photosynthetic organisms have existed on Earth for at least 2×10^9 years. The total amount of organic compounds formed each year by photosynthesis is about 1.0×10^{17} grams or a trillion metric tons. The total mass of organic material produced by green plants during the biological history of the Earth has been estimated to be about 6×10^{25} grams. This is an enormous weight when compared to the mass of the Earth, as it represents one percent of the Earth's mass of 6×10^{27} grams.

A biochemical operation of this magnitude would quickly deplete the Earth's atmosphere of its carbon dioxide. Since our atmosphere still contains CO_2, it must be returned to the atmosphere by equally large-scale processes. That is, the rate of CO_2 consumption in photosynthesis is just about balanced by the rate of restoration. This relationship between CO_2 uptake and O_2 in photosynthesis is illustrated in Figure 5.2 in which the spectrum for the uptake of CO_2 in photosynthesis corresponds to the absorption spectral peaks for chlorophyll. Respiration by plants

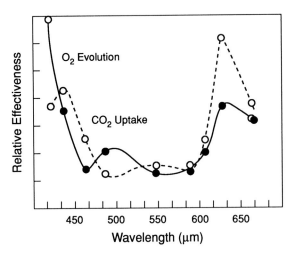

FIGURE 5.2   Carbon dioxide uptake and oxygen evolution during photosynthesis.

and animals is responsible for this restoration of atmospheric CO_2 in accordance
with the equation:

$$(CH_2O)_n + O_2 \rightarrow CO_2 + H_2O$$

organic
compounds

This restoration is supplemented by the decay of organic matter and the burning of
fossil fuels, both of which yield CO_2. As a result, the average surface concentration of CO_2 in the air is about 0.03% and has remained practically constant for
thousands of years. The set of processes maintaining the CO_2 balance, known as
the *carbon cycle,* is important not only in maintaining a constant chemical atmospheric environment but also in regulating the surface temperature of the Earth,
since CO_2 absorbs infrared energy. Animals consume carbohydrates and utilize O_2
as an electron acceptor, so the carbon cycle ultimately results in the conversion of
light energy into useful biological chemical energy, using electrons as energy
transducers.

Photosynthesis is equally important for regulating the oxygen content of the
atmosphere. If too much oxygen had been created too early, the greenhouse effect
would have been lost and the global temperature would have plunged. If oxygen
production had proceeded too slowly, the greenhouse effect would have trapped
too much heat. Therefore, biological and other mechanisms were at work to
modulate the atmospheric oxygen, since there was very little, if any, oxygen in the
atmosphere. The first organisms that evolved did not produce O_2; only later when
the photosynthesizing blue-green algae (cyanobacteria) emerged did O_2 enter the
atmosphere. It has been estimated that the current rate of photosynthesis produces

almost the entire oxygen content of the atmosphere necessary for the respiration of all animals. The oxygen concentration has been maintained at its present level of around 20% for a very long period of time as a result of a state of balance in the oxygen cycle. The emergence of photosynthesis, then, has played a crucial role in the evolution of life on Earth.

The beginning of our understanding of photosynthesis goes back to the 1770s when Lavoisier (1774) in France determined the composition of air and Joseph Priestley (1772) in England discovered oxygen. Priestley then showed that oxygen was produced by algae and green plants. The Dutch physician Jan Ingen-Housz observed in 1779 that in the process of plant respiration, green plants at night or in the dark gave off a "dangerous air," carbon dioxide, which was purified by sunlight. The Swiss clergyman Jan Senebier in 1782 was able to demonstrate that carbon dioxide was absorbed by the leaves of green plants and, when exposed to sunlight, oxygen evolved. In 1798, Ingen-Housz published his theory that carbon, already recognized at the time as an important element in the composition of organic molecules, was derived from carbon dioxide during photosynthesis. It was another Swiss scientist, Nicholas de Saussure (1804), who surmised that water played an essential role, and the picture changed to one in which light acted on both carbon dioxide and water. Pelletier and Caventou in 1818 identified chlorophyll as the green pigment of plant tissue and by 1837 chlorophyll was identified with the chloroplasts of plant cells. However, the crux of the phenomenon of photosynthesis was visualized in 1845 by Robert Mayer, a German physicist and physician, who pointed out that the photosynthetic process was the conversion of light energy to chemical energy. By 1882, Englemann had demonstrated that the site of photosynthesis resided within the chloroplast of plant cells and that upon light absorption oxygen was liberated. From then on meaningful research to understand the mechanisms of photosynthesis had begun. An account of these early investigations into photosynthesis as it developed is summarized by Rabinowitch in the three volumes of *Photosynthesis and Related Processes* (1945, 1951, 1956).

COMPARATIVE ASPECTS OF PHOTOSYNTHESIS

Originally, early anaerobic bacteria were not able to photosynthesize but were able to utilize inorganic compounds such as hydrogen sulfide and hydrogen gas, as well as organic compounds found in their environment, as a source of energy. In time, some of these bacteria developed metabolic pathways that led to the synthesis of porphyrins, bacterial chlorophyll, and hence to the evolution of bacterial photosynthesis. When photosynthetic bacteria became well established, a second kind of photosynthesis became possible in which a more prevalent source of electrons was formed by the oxidation of water molecules. As a result, free oxygen entered the atmosphere and became available for further chemical synthesis.

For anaerobic bacteria, oxygen was a deadly poison, but as the population capable of photosynthesis increased, so did the oxygen content of the oceans and the atmosphere. It has been hypothesized that three billion years ago the level of atmospheric oxygen was less than 0.001% of the present level and by one billion years ago had probably increased to about 0.1% of the present level.

The first organisms on Earth that adapted to the presence of free oxygen led to the appearance of "blue-green" algae, now known as cyanobacteria. Compared with algae and green plants, cyanobacteria are primitive. Their success was manifest in the evolution of new mutants of bacteria that utilized little oxygen in their metabolic processes. Van Niel (1941, 1943) pointed out that the photosynthetic cyanobacteria and the purple bacteria represent remnants of what was originally a much wider class of organisms having a photosynthetic system simpler than that of green plants. The metabolism of purple bacteria could serve as an example of the kind of photochemistry which may have preceded that of the green plants on the evolutionary time scale. These bacteria cannot evolve oxygen, though many tolerate oxygen. Because they require energy-rich hydrogen donors (H_2S, etc.) to reduce CO_2, they do not contribute much to the store of free energy in the living world. In photosynthetic bacteria, bacteriochlorophyll is responsible for the utilization of light energy and is similar in structure to the chlorophyll molecule of green plants (Figure 3.4).

Let us briefly examine some aspects of the comparative biochemistry of photosynthesis. Organisms are classified as *autotrophic,* those which obtain their energy for growth from sources other than organic molecules; *chemoautotrophic,* those which obtain their energy from oxidizable inorganic chemicals; and *photoautotrophic,* those which obtain their energy directly from light. The photoautotrophic fall into three separate groups: green plants, pigmented sulfur bacteria, and pigmented non-sulfur bacteria. Their light-driven reactions can be expressed by the following chemical equations:

$$\text{Green plants: } CO_2 + H_2O \xrightarrow{\text{light}} \underset{\text{Organic matter}}{(CH_2O)} + O_2 \uparrow$$

$$\text{Sulfur bacteria: } CO_2 + H_2S \xrightarrow{\text{light}} (CH_2O) + S$$

$$\text{Nonsulfur bacteria: } CO_2 + \text{succinate} \xrightarrow{\text{light}} (CH_2O) + \text{fumarate}$$

For the sulfur bacteria, H_2S can be replaced by $Na_2S_2O_3$, $Na_2Se_4O_7 - H_2SO_4$, or H_2Se. For non-sulfur bacteria, the organic donor succinate can be replaced by many different organic acids that have two electrons to spare, as is generalized by:

$$CO_2 + H_2A \xrightarrow{\text{light}} (CH_2O) + A$$

The characterization of the various types of photosynthesis led to a generalization by Van Niel (1941, 1949) of what is known as the "comparative biochemistry of photosynthesis." Actually, the photosynthetic reaction can be schematized even further without involving carbon dioxide, for in the Hill reaction (production of O_2 by isolated chloroplasts in light), a quinone or ferric ion can accept the hydrogens that are activated by the light reaction, so the general formula can be written:

$$B + H_2A \xrightarrow{\text{light}} BH_2 + A$$

or simply as light-induced oxidation-reduction reaction.

Many researchers attacking the mechanisms of photosynthesis hypothesized that the photochemical reaction in green plant photosynthesis is a photolysis (light-induced decomposition) of water. Indeed, experimental evidence has shown that water is photolyzed in both plant and bacterial photosynthesis. Therefore, a more acceptable equation for the bacterial systems would include both water and a general hydrogen donor. It would also exclude oxygen evolution. For example:

$$CO_2 + 2H_2A + H_2O \xrightarrow{\text{light}} (CH_2O) + 2H_2O + 2A$$

A simple formulation for the mechanism of the overall process has yet to be conceived. Some aspects, however, are reasonably well understood. The general photosynthetic light reaction can be represented schematically as:

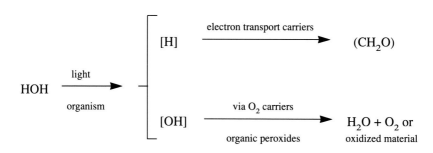

The photolysis of water can be regarded, then, as the major achievement of the chlorophyll system.

CAROTENOIDS IN PHOTOSYNTHESIS

We can now ask what function the carotenoids play in the primary photosynthetic reactions. Experimental investigations support the view that the carotenoids participate directly in the primary act of photosynthesis. This is now being examined by employing the method of molecular genetics in bacterial and plant photosynthetic systems. Anaerobic bacterial photosynthetic systems are unique in that they do not evolve oxygen. *Rhodopseudomonas spheroides* blue-green mutant lacks carotenoid pigments present in the wild type. This mutant has been extensively studied and found to grow well photosynthetically without these carotenoids that were considered necessary so long as the growth medium was free of oxygen (Stanier, 1959). When exposed to light and oxygen, however, rapid death and bacteriochlorophyll decomposition occurred. This demonstrated that the light-trapping process was not dependent on the carotenoids if oxygen was absent and indicated that bacteriochlorophyll, and hence chlorophyll, was the primary pigment necessary for the light reaction. These experiments also indicated that the carotenoids were not essential, except in catalytic amounts, for green plant photosynthesis but were necessary for protection against photodynamic destruction.

Action spectra have shown a direct relationship between the synthetic pathways of chlorophyll and carotenoid synthesis in algae and plants (Wolken, 1967; Ogawa et al., 1973). The relationship between chlorophyll and carotenoid pigments is probably that the C_{20} phytol chain of chlorophyll is derived from precursors of C_{40} carotenoids. The *Rhodopseudomonas spheroides* blue-green mutant does not have the C_{40} carotenoid precursor, phytoene, which can give rise to phytol via a divergent pathway. The carotenoid synthetic pathway beyond phytoene is genetically blocked in the mutant. One hypothesis is that the carotenoids combine with the oxidized portion of the photosynthetically split water molecule by forming epoxides across the numerous double bonds, with one or more epoxide group resulting:

$$-C=C-C=C-C=C-$$

$$-\underset{\underset{O}{\diagdown\diagup}}{C-C}-\underset{\underset{O}{\diagdown\diagup}}{C-C}-\underset{\underset{O}{\diagdown\diagup}}{C-C}-$$

Such epoxide formation has been demonstrated in leaves *in vitro*. Therefore the carotenoids can act as a filter to screen the light and prevent photo-oxidation (photodestruction) of chlorophyll at high light intensities. On the other hand, they could function as an accessory pigment molecule in the energy transfer process.

THE MECHANISM

In examining the photosynthetic mechanisms, two distinct but related processes occur. One process is the biochemical conversion of carbon dioxide to carbohydrates and inorganic compounds that were indicated by the equation for green plants:

$$CO_2 + H_2O \quad \xrightarrow[\text{chlorophyll}]{\text{light}} \quad (CH_2O)_n + O_2$$
organic
compounds

The other process is the photophysics, the transduction of light energy to chemical energy, which is the more difficult to understand. Thus, photosynthesis is a series of light and dark reactions. The reduction of carbon dioxide to carbohydrates is a dark reaction and is separate from the primary light quantum conversion.

We can say, then, that light energy is converted into chemical energy to form carbohydrates and oxygen. In this process, the light energy is absorbed by chlorophyll and related pigments and is converted into chemical potential energy in the form of certain compounds. These compounds then react with water, liberating oxygen and reducing agents and other cofactors which contain high chemical potential energy. Finally, these reducing and energetic cofactors react with carbon dioxide and other inorganic compounds to produce organic compounds.

Melvin Calvin and his associates, in the 1940s at the University of California at Berkeley, began to study the pathway of carbon reduction during photosynthesis using ^{14}C. They identified phosphoglyceric acid as the first stable product of carbon reduction during photosynthesis. With the developments of two-dimensional paper chromatography and radioautography, analytical tools were available for separating and detecting minute amounts of radioactive compounds formed in the plant during photosynthesis. By these methods, the intermediates in the carbon reduction cycle were found to be sugar phosphates. Calvin (1962) confirmed Blackman's hypothesis that light was necessary for only two processes: (1) to produce ribulose diphosphate, the acceptor of CO_2, by the phosphorylation of ribose monophosphate; and (2) to permit the reduction of carboxyl groups of phosphoglyceric acid (PGA) in the aldehyde group by the intermediate 1,3-diphosphoglyceric acid. In these reactions, the donor of the phosphoryl group is adenosine triphosphate (ATP). The process of oxidative phosphorylation is linked to the synthesis of ATP, the energy-rich storing mechanism in the energetics of all life processes.

ENZYMES

Researchers have looked for the participation of a specific enzyme system in photosynthesis for capturing electrons. Hill and Bendall (1960) demonstrated experimentally that a cytochrome, or a cytochrome system, is coupled to the chlorophyll–protein complex in the chloroplast and that it functions as an electron carrier during the initial reactions of the photochemical process. As Hill points out, mitochondria and chloroplasts show close resemblance with respect to the structure-bound cytochromes. This indicates that the chloroplasts may belong to the same category as the mitochondria, with cytochrome *a* being replaced by chlorophyll. There is also the possibility that the photochemistry initiated by light absorption in photosynthesis involves the cytochrome directly. For example, in *Euglena,* two spectrally different cytochromes have been isolated. One, from the light-grown photosynthetic cells, is designated as "cytochrome-552" (a c-type cytochrome). In its reduced state, its absorption peaks are at 552, 523, and 416 nm (Figure 3.3). The other cytochrome, isolated from dark brown *Euglena,* is referred to as "cytochrome-556," because in the reduced state it has absorption peaks 556, 525, and 412 nm. Its spectrum is close to that of cytochrome *f,* which is associated with green plants. The ratio of chlorophyll *a* to cytochrome-552 is approximately 300:1.

Similarities in the photosynthetic systems of algae and green plants have led to the suggestion that the cyanobacteria became a symbiot of an early cell. Thus the chloroplast ferredoxins are derived from a common ancestor. Arnon (1965) has shown that ferredoxin is a key photochemical component of the process of photosynthesis by chloroplasts. Ferredoxins are biological reducing agents. They transfer electrons, and as such they participate in metabolic processes as diverse as the fixation of atmospheric nitrogen, the production of hydrogen, and photosynthesis. Ferredoxins are iron-protein molecules that are comprised of about 50 to 100 amino acids and inorganic sulphur (Figure 5.3). They are relatively small protein molecules (Table 5.1) in comparison to most proteins. Ferredoxin is not a heme protein like cytochromes and, unlike the cytochromes which exhibit well-defined absorption peaks in the reduced state, have distinct absorption peaks in the oxidized state (Figure 5.4). According to Arnon, the photoreduction of ferredoxin is coupled with oxygen evolution and with photosynthetic phosphorylation. It is interesting to note that the ratio of chlorophyll to ferredoxin molecules is of the order of 300:1, like that of cytochrome.

Plastoquinone is another important intermediate in photosynthesis and is found in cyanobacteria, in green, red, and brown algae, and in green plant chloroplasts. The ratio of the total various quinones to chlorophyll was found to be about 150:1 (Amesz, 1973).

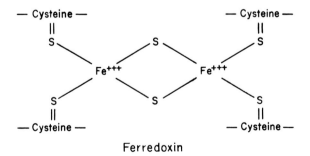

Ferredoxin

FIGURE 5.3   Basic chemical structure of Ferredoxin.

TABLE 5.1 Some Comparative Properties of Ferredoxin[a]

	Bacteria	Photosynthetic bacteria	Plants
	Clostridium pasteurianum	*Chromatium*	Spinach
Iron content (atoms/molecule protein)	7	3	2
Inorganic sulfide (moles/molecule protein)	7	3	2
Redox potential (volts, pH 7.55)	−.417	−.490	−.432
Molecular weight[+]	6,000	6,000	13,000

[a]Experimental data from D.I. Arnon (1965, p. 1464) and other sources.

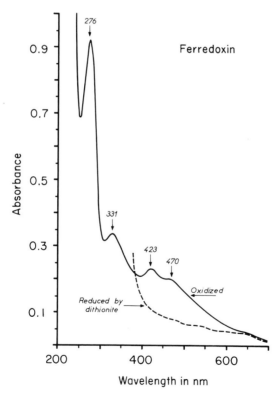

FIGURE 5.4   Absorption spectra for oxidized and reduced (----) ferredoxin. Extracted from spinach.

TWO PHOTOSYSTEMS IN PHOTOSYNTHESIS

Experimental studies by Emerson and Lewis (1943) indicated that the photo-synthetic process was not a simple photoreception sensitized by chlorophyll. They observed that the quantum yield in photosynthesis was constant between 500 and 680 nm but dropped dramatically beyond 680 nm. Since chlorophyll *a* is the major light absorber in this region of the spectrum (Figures 3.5 and 5.5), it seemed that light absorption by chlorophyll *a* alone was not sufficient for photosynthesis to proceed. Therefore, it was assumed that there are forms of chlorophyll *a* in the living cell which differ in the way they are complexed with their proteins, or perhaps they are associated with accessory pigments.

Emerson later (1956), using *Chlorella* and monochromatic light, observed that the low efficiency of photosynthesis beyond 680 nm could be considerably improved by simultaneous illumination with a shorter wavelength, of blue light at 480 nm. Thus, the low efficiency of absorption in the far-red beyond 680 nm would require another pigment-complex absorbing below 680 nm. To account for these experimental results, Duysens (1964) postulated that two pigment photosystems for photosynthesis are required. They are known as Photosystem I and Photosystem II. More recent studies have been made to clarify how these two photosystems function in photosynthesis.

A scheme to illustrate this complex electron transport chain is depicted in Figure 5.6. In this scheme, water serves as the electron donor in a photoreaction promoted by a chlorophyll complex, P680, referred to as Photosystem II. The electron

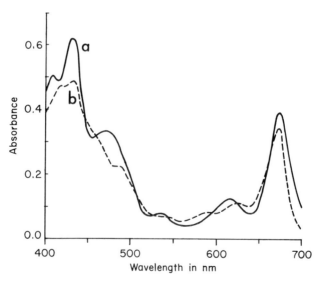

FIGURE 5.5   (a) Absorption spectrum of *Euglena gracilis* chloroplast (obtained by microspectrophotometry) compared to (b) absorption spectrum of chlorophyll *a* (in acetone).

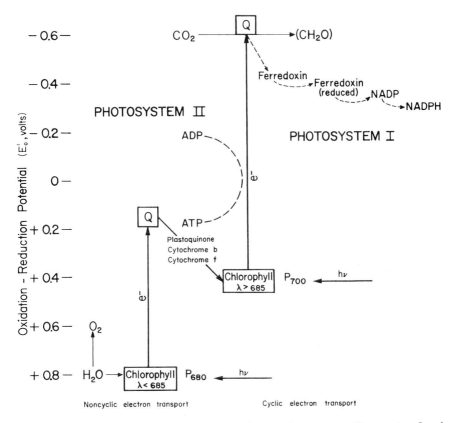

FIGURE 5.6   Schematic diagram of the two pigment photosystems, Photosystem I and Photosystem II, electron transport chain in photosynthesis.

acceptor molecule is probably a quinone Q that has a redox potential around 0.0 to $+0.18$ V. The reducer Q transfers its electron to Photosystem I through a series of compounds, including plastoquinone and several cytochromes. This chlorophyll complex, containing mainly chlorophyll a and absorbing at longer wavelengths, is called P700 because it behaves differently from a typical chlorophyll, and it has a redox potential of $+0.4$ V. Thus, an electron moving between the two photosystems loses the equivalent potential of about 0.2 to 0.4 V. This is enough energy to promote the formation of one or two ATP molecules from adenosine diphosphate (ADP), and inorganic phosphate. In Photosystem I, the primary electron acceptor is quinone. The light absorbed by Photosystem I is then used to reduce ferredoxin. The final product of this electron transport scheme is nicotinamide adenine dinucleotide phosphate hydride (NADPH). Therefore, the two photochemical, oxidation-reduction reactions are driven by two photopigment systems, Photosystem I and Photosystem II, and provide the high-energy phosphates (reduced NADPH and ATP) needed for the synthesis of carbohydrates and proteins from CO_2 and water.

> *From the simplest substances, carbon dioxide, water, and sunlight, auto-trophic plants produce enormous quantities of organic matter. . . . Synthesis of all this diverse vegetable material hinges upon photochemical reactions that take place within the green parts of plants.*
> —H. H. STRAIN, 1944, *Annual Review of Biochemistry*

THE CHLOROPLAST

The "eyes" of plants are chloroplasts; they are the photoreceptors that initiate the photoprocesses of photosynthesis upon light absorption.

The question arises: How are chloroplasts structured at the molecular level to capture the energies of light and to convert the energy to chemical energy? In unraveling the molecular structure and chemistry, insight into how the chloroplast functions in photosynthesis has been achieved. Chemical analysis of chloroplasts isolated from a variety of plant species shows that chlorophylls (5% to 10%) and the other major constituents are proteins (35% to 55%), lipids (18% to 37%), carotenoids (2%), inorganic matter (5% to 8% on a dry weight basis), and the nucleic acids RNA and DNA (1% to 3%). All chloroplasts, except for bacterial chromatophores, contain chlorophyll *a*, and all higher plants and green algae contain in addition chlorophyll *b* and the carotenoid β-carotene. The total number of chlorophyll molecules per chloroplast is of the order of 1.0×10^9 (Table 5.2).

TABLE 5.2   Chlorophyll Concentration in Chloroplasts

Organism	Volume of chloroplast (ml)	Chlorophyll molecules per chloroplast	Concentration of chlorophyll (moles/liter)
Elodea densa (green plant)	2.8×10^{-11}	1.7×10^9	0.100
Mnium (moss)	4.1×10^{-11}	1.6×10^9	0.065
Euglena gracilis (algae flagellate)	6.65×10^{-11}	$1.02 \times 10^{9*}$	0.025
Poteriochromonas stipitata (crysomonad)	1.1×10^{-11}	0.11×10^9	0.016

*Number of chlorophyll molecules 1.02×10^9 ($0.88-1.36 \times 10^9$) (calculated from chloroplast extract in solution) 1.34×10^9 (calculated from a single chloroplast using mirospectrophotometry).
From Wolken (1975).

The fact that DNA is present in chloroplasts indicates that they possess an autonomous genetic system different from the cell itself. The amount of DNA in chloroplasts is about the same as in *Escherichia coli* and possesses sufficient genetic information for a large number of physiological functions. Chloroplast DNA has a nucleotide composition sharply different from that of the nuclear DNA (Brawerman and Eisenstadt, 1964). Chloroplasts also contain messenger RNA in sufficient quantity for maximum activity of their protein-synthesizing system. A mechanism could be postulated by assuming that the messenger RNA molecules for the structural proteins of the chloroplast are generated *in situ* by the chloroplast DNA. The replication and turnover of chloroplast DNA in *Euglena* have been shown to be more rapid than those of nuclear DNA (Manning and Richards, 1972). This information has raised many interesting questions as to the origin of chloroplasts in plant cells (Cohen, 1970, 1973; Sager, 1972; Margulis, 1970, 1982).

The chloroplast molecular structure

How is the chloroplast structured to function in photosynthesis? The chloroplasts in photosynthetic bacteria are described as *chromatophores,* in algae as *plastids,* and in all green plants as chloroplasts. The chloroplasts of algae and green plants are of various shapes, but generally they are ellipsoid bodies from 1 to 5 μm in diameter and from 1 to 10 μm in length. Chloroplasts observed with the polarizing microscope show both form and intrinsic birefringence. With the fluorescence microscope they show measurable fluorescence. These observations indicate that chloroplasts possess a highly ordered molecular structure.

Electron microscopy of chloroplasts in a variety of plant cells reveals that they consist of membranes, as can be seen in a section through the *Euglena* chloroplast and the green plant *Elodea* chloroplast (Figures 5.7, 5.8); they are structured of regularly spaced membranes, or lamellae. Green plant chloroplasts contain *grana* that form closed, flattened vesicles or discs; these membrane structures are referred to as *thylakoids.* The chloroplast membranes are bilayers of lipids and proteins to which chlorophyll is associated. The intermembrane spaces contain water, enzymes, and dissolved salts.

The number of chlorophyll molecules per chloroplast, from photosynthetic bacteria to higher plants, is of the order of 10^9 molecules. The number of chlorophyll molecules in the chloroplast is directly related to the number of membrane surfaces. This suggests a mode of growth regulation on the molecular level for chloroplast development. That is, the chlorophyll molecules would be spread as monolayers on the surfaces of the chloroplast membranes as depicted in the molecular model (Figure 5.9). This maximizes the surface area of the chlorophyll molecule for light absorption and for energy transfer at specific sites on the membrane. Such a highly ordered membrane structure provides not only for the energetic interaction of the chlorophyll and carotenoid molecules but also reactive sites for the necessary enzymatic reactions.

To establish that the chlorophyll molecules are spread as a monomolecular layer

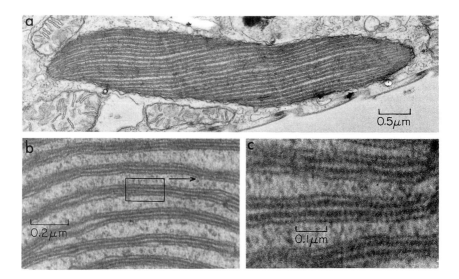

FIGURE 5.7   The chloroplast plastid structure of *Euglena gracilis* (*a*). Chloroplast membranes at higher magnification (*b*) and at greater resolution (*c*). Electron micrographs.

on the membrane surfaces, the cross-sectional area available for the porphyrin part of the chlorophyll molecule was calculated (Wolken, 1975). To do this, the geometry of an individual chloroplast (its length, diameter, and number of membrane surfaces) was measured from numerous electron micrographs. The calculated cross-sectional area of the chlorophyll molecule was found to be 222 Å^2 for the *Euglena* chloroplast; the cross-sectional area in a variety of plant chloroplasts was found to be around 200 Å^2. These calculations correspond well with the cross-

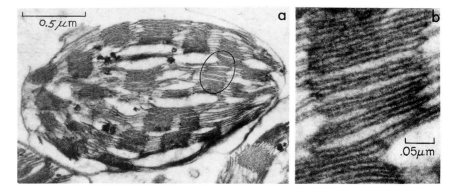

FIGURE 5.8   The chloroplast structure of the plant *Elodea densa* consisting of grana (*a*) and enlarged membranes of a granum (*b*). Electron micrographs.

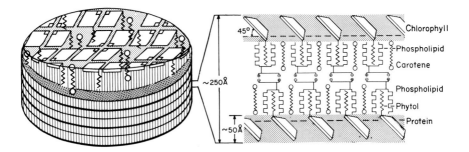

FIGURE 5.9   Schematic molecular structural model of a chloroplast.

sectional area measurements of a porphyrin molecule when spread on a water-air interface.

In the molecular model (Figure 5.9), it was assumed that the chlorophyll molecules are oriented as a monolayer on the surfaces of the lipoprotein membranes. The chloroplast lamellar network shows that four chlorophyll molecules are united to form tetrads and are oriented so that only one of the phytol tails of chlorophyll is located at each intersection of the rectangular network (Figure 5.9). This arrangement has the advantage of leaving adequate space for at least one carotenoid molecule for every three chlorophyll molecules. Since the molecular weight of the carotenoid molecules is one-half to two-thirds the molecular weight of the chlorophyll molecules, a weight ratio of chlorophyll to carotenoid of approximately 4:1 to 6:1 would be expected. On the other hand, the carotenoid molecules are slender, linear molecules, about 5Å in diameter, and therefore more than one molecule could conveniently fit into the 15Å × 15Å hole formed by the chlorophyll tetrads. From symmetry one might expect as many as four molecules per hole, but this would lead to a very tight, energetically improbable fitting. One can therefore put a lower limit on the number of chlorophyll to carotenoid molecules of roughly 1:1 and a weight ratio of 2:1.

In Table 5.3, we see that the mono- and digalactosyl diglycerides account for the major lipids in the chloroplast. These lipids, because of their properties, can form a lipid or lipoprotein matrix for the chlorophyll monolayers. From spatial considerations the ratio of two galacotsyl diglyceride molecules to one chlorophyll molecule could stabilize all the chlorophyll molecules in the monolayer. That is, there would be one phytol chain of chlorophyll for four *cis*-unsaturated acyl chains of galactosyl diglyceride (Rosenberg, 1967). Such a relationship fits with the molecular model for chlorophyll in the chloroplast membrane. There are, of course, other possible ways in which the chlorophyll molecules could be oriented in the chloroplast membranes. If the porphyrin parts of the chlorophyll molecule lie at 0° as depicted in Figure 5.9, their greatest cross-section would be available. If they are oriented at increasing angles to 45°, the cross-sectional area would be decreased to about 100 Å^2. Since the chlorophyll molecules in the chloroplast are

TABLE 5.3   Chemical Analysis of Spinach Chloroplast
Quantasome Chlorophylls and Lipids

230	Chlorophylls	
	160	Chlorophyll *a*
	70	Chlorophyll *b*
48	Carotenoids	
	14	β-carotene
	Lipids	
	116	Phospholipids
	48	Sulpholipids
	114	Digalactosyl diglyceride
	346	Monogalactosyl diglyceride

Data taken in part from Park and Biggins (1964, p. 1010) and Wolken (1975).

in a dynamic state, they would orient themselves to maximize their largest available cross-section for light capture.

The photoreceptor cell membrane of halophilic bacteria

An important goal in membrane biology and biophysics is to determine how the membrane proteins function both as energy transducers of light and in active ion transport. A highly specialized photoreceptor membrane system is that of the bacterium *Halobacterium halobium*. These halophilic (salt loving) bacteria are extremely interesting, for they live and grow in high salt concentrations (25% NaCl) at temperatures near 44°C and in direct sunlight. The bacterium *Halobacterium halobium* has surprisingly incorporated in its cell "purple membrane" the visual pigment, a rhodopsin, chemically identified as bacteriorhodopsin. The function of bacteriorhodopsin in these bacteria is not for visual excitation, as in the retinal photoreceptors of the eye, but bacteriorhodopsin functions in the photochemical processes of photophosphorylation, as does chlorophyll in the energetics of photosynthesis.

Bacteriorhodopsin is an integral protein component of the purple membrane that forms two-dimensional crystals in the lipid bilayers of the membrane. In its native form, three bacteriorhodopsin molecules form a trimer, each separated by 20 Å. The trimers are located in a two-dimensional hexagonal lattice in the purple membrane with a lattice constant of 67 Å. A molecular structural model of bacteriorhodopsin in the membrane was developed by Unwin and Henderson (1975). In their model the protein bacteriorhodopsin comprises seven rod segments as α-helices, which are membrane spanning and oriented perpendicularly to the plane of the lipid bilayer. This is schematized in the retinal rod membranes of the vertebrate eye (Figure 9.11).

In unraveling the chemistry and molecular structure of the photosynthetic mem-branes, much insight into the mechanism of photosynthesis has been achieved. In reviewing photosynthesis, it is clear that many fascinating and perplexing prob-lems remain to be solved. How did the first photosynthetic system evolve? How precisely do the physical and chemical reactions of these photosystems function? Finally, when these are understood, how can we reproduce photosynthesis outside of the living cell?

In searching for how the photoreceptor system evolved, clues can be found among unicellular organisms. These organisms long ago developed photoreceptors for light searching as exhibited in phototropism and phototaxis. Such phototactic behavior by relatively simple unicellular organisms leads us to examine their photosensory systems in order to find out whether there are in fact common physical-chemical relationships to the photosensory mechanisms of more highly evolved animals.

CHAPTER SIX

Phototactic Behavior: Searching for Light

> *The ability to respond to stimuli is one of the characteristics of living things which appear early in the course of evolution, almost as soon as the aggregation of large molecules to form a cell.*
>
> —E. NEWTON HARVEY, 1960, *Comparative Biochemistry*

> *How a nerve comes to be sensitive to light, hardly concerns us more than how life itself originated; but I may remark that, some of the lowest organisms, in which nerves cannot be detected, are capable of perceiving light.*
>
> —CHARLES DARWIN, 1859, *Origin of Species*

All organisms respond in some way to light; they move, orient, swim, or fly to or away from the light—they are phototactic. Organisms have evolved photoreceptors for detecting light, measuring its intensity, and selecting the wavelengths of light for function—all in search of an optimal environment for survival.

Phototactic behavioral responses are described as *phototropism* and *phototaxis*. Phototropism, as defined here, is simply a positive or negative orientation of a part or of the whole organism to move toward or away from light. The bending and twisting of a plant leaf to present its surface to the light is an example of positive phototropism. This ability of a plant to orient in search of light increases its efficiency for photosynthesis, which in effect determines the plant's biochemistry and growth.

Julius von Sachs in 1864 observed that the bending of plants toward light is stimulated primarily by blue light. Experimental studies since then have shown that blue light induces phototropism in fungi, ferns, and higher plants. In higher plants, both blue and red light are effective, and for mosses only red light is effective. To produce the response, light must be absorbed. This means that plants exhibiting phototropic behavior contain blue-absorbing and red-absorbing pigments.

Researchers have turned to investigating the photobehavioral mechanisms phototropism and phototaxis in algae, bacteria, fungi, and protozoa. These unicellular organisms are the simplest for quantitative studies of the relationship between light and behavioral responses. The patterns of phototactic movement in response to light stimuli bear directly on the underlying photoreceptor mechanisms of more highly evolved plant and animal sensory systems. Therefore, isolating the function of light in phototactic behavior is of considerable interest.

PHOTOTROPISM

To begin the search for the mechanisms that detect and respond to light, I have turned to fungi, which are widespread in nature. Many fungi are light-seeking; they bend and turn toward the light—they are phototropic. Fungi do not synthesize chlorophyll and are not capable of photosynthesis. They use light as a signal, and their phototropic behavior can be isolated without having to deal with the complexity of photosynthesis. Experimental studies with the well-studied fungus *Phycomyces blakesleeanus* were pursued in search of the phototropic detecting photoreceptor system. The growth, development, and photosensory phototropic behavior of *Phycomyces* are described in the reviews of Shropshire (1963) and Bergman et al. (1969). More recent experimental studies are found in Delbrück (1976), Lipson (1983), Wolken (1975, 1986), and in the review *Phycomyces* (edited by Cerdà-Olmeda and Lipson; 1987).

Phycomyces is a single aerial cell that matures in distinct stages designated as I–IVb. In the growth process, the sporangiophore (fruiting body) grows to more than 10 cm in length and to about 0.1 mm in diameter (Figure 6.1). The growth stages are related to the elongation of the sporangiophore and the development of the sporangium, or spore head. The general appearance of the sporangiophore in Stage IVb is that of a nearly transparent, cylindrical filament supporting a spherical black sporangium. The growth time from Stage I to Stage IVb is approximately 100 hours; it is dependent on the light, temperature, and humidity of the growth medium.

The sporangiophore exhibits two light-sensitive responses: that of light growth and that of phototropism (Figure 6.1). Light functions as a signal to alter the growth rate in space and time. The sporangiophore exhibits phototropism when it is unilaterally illuminated with light. It is a light-searching organism that tracks the direction of light (Figure 6.2). *Phycomyces* is positively phototropic to visible wavelengths from 300 to 510 nm, and its maximum response is in the blue. It is negatively phototropic to ultraviolet wavelengths shorter than 300 nm.

The sporangiophore is also sensitive to gravity, touch, and the presence of

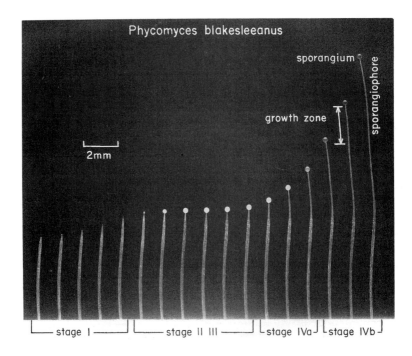

FIGURE 6.1   *Phycomyces blakesleeanus* during growth in time, from Stages I–IVb.

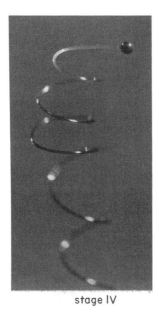

stage IV

FIGURE 6.2   The response of *Phycomyces blakesleeanus*' response to a moving beam of light. Demonstration of helical phototropism, Stage IV. (Courtesy of Professor David Dennison, Dartmouth College, Hanover, New Hampshire.)

nearby objects. Why is all this sensory behavior built into such a primitive organism? Since it can so keenly sense its environment, what can be learned from it about more highly evolved plant and animal photosensory systems? As an organism, it possesses no obvious structure that resembles an eye or a nervous system. What are the photoreceptor molecules and receptor structures that permit the organism to respond to environmental stimuli, behavior we normally associate with animals?

In search of answers to these questions, *Phycomyces'* structure and pigments were experimentally determined. Mutants can be produced by physical and chemical means that are genetically altered. These mutants can be selected and are (1) fully sensitive to light, (2) "night-blind" or sensitive to only high light intensity (as are the cones in the retina of vertebrate eyes), or (3) insensitive to light and considered "blind." If one can establish the genetic basis for the genes that determine the molecular basis for the ability to "see" or for "blindness," then the genes controlling the biosynthesis of their pigment photoreceptor molecules can be identified by methods of molecular genetics.

In search of the photoreceptor pigment molecule

The *Phycomyces* photosensory region is the light growth-zone of the sporangiophore. This zone extends from 0.1 mm to about 3 mm below the sporangium and occupies a surface area of about 6×10^{-3} cm^2. The sporangiophore is most photosensitive in Stage I and in Stage IVb. The action spectra for *Phycomyces* phototropism shows absorption peaks around 280, 365 to 385, 420 to 425, and 445 to 485 nm (Delbrück and Shropshire, 1960; Curry and Thimann, 1961; Galland and Lipson, 1985). So we infer that the photoreceptor molecule should have similar absorption peaks. The absorption spectrum obtained by microspectrophotometry for Stage IVb of the wild-type sporangiophore through the growth-zone is shown in Figure 6.3. In scanning down the sporangiophore from 0.1 mm to 2 mm below the sporangium, the absorption spectrum gradually shifts from that of Figure 6.3a to that of 6.3b. The spectral absorption peaks at 430, 460, and 480 nm in Figure 6.3a comprise the absorption spectrum of β-carotene and correspond to those in the visible part of the action spectrum for *Phycomyces* phototropism. The absorption spectrum of the albino mutant (Figure 6.3c), in the ultraviolet near 280 and 370 nm, and in the visible around 450 nm, are similar to those of a flavin or flavoprotein (Figure 6.17c). It is interesting to note that all absorption peaks *together* at 280, 370, 435, 460, and 485 nm correspond to the action spectrum for phototropism (Figure 6.3).

Phycomyces "albino" mutants are deficient in β-carotene but are still phototropic, so their absorption spectra should be more informative as to their photoreceptor pigment. The absorption spectrum within the growth-zone of the albino mutant has absorption peaks near 230, 267, and 370 nm, that of a flavin. Neither the action spectra nor the absorption spectrum of the phototropic growth-zone of *Phycomyces* has been sufficiently informative to clearly identify the primary pho-

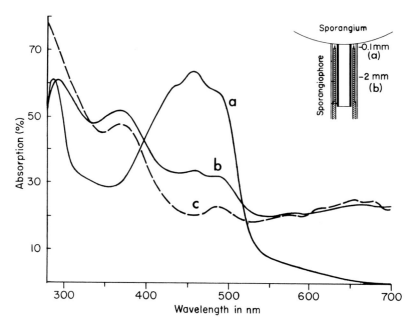

FIGURE 6.3 *Phycomyces* absorption spectra in the light growth zone. (a) Wild-type, Stage IVb, at 0.1 mm to 2 mm; (b) from 2 mm and below the sporangiophore; (c) compared to albino mutant absorption spectrum in the light growth zone.

toreceptor molecule. It is difficult to distinguish between carotenoids and flavins, since they closely resemble the absorption spectral peaks in the visible region, and even their fluorescence is inconclusive.

It therefore was of interest to determine whether flavins were present and in what concentration. From phycosporangiophores flavins were extracted, and the total number of flavin molecules for the wild type was 13×10^{12} and that for the albino mutant was 5×10^{12}. The flavins identified were riboflavin, lumi-flavin, lumichrome, flavin adenine dinucleotide (FAD), and flavin mononucleotide (FMN). How reasonable are these findings about the sporangiophore photorecep-tor pigment? The microspectrophotometry of the growth-zone showed a shift in a spectral characteristic from a typical carotenoid spectrum (Figure 6.3a) to a typical reduced flavin or semiquinone spectrum (Figure 6.3b). These spectra would sug-gest that a flavoprotein is the photoreceptor molecule for phototropism (Delbrück, 1976; Wolken, 1972, 1975). Nevertheless, it was still suspected that the photo-receptor molecule could be a carotenoid and that retinal, the chromophore of the visual pigment rhodopsin, would be found. In *Phycomyces* wild type, small quan-tities of retinal were extracted from the sporangiophores (Meissner and Delbrück, 1968), but at that time it was ruled out since retinal was not detectable in β-carotene-deficient *Phycomyces* mutants, which were still phototropic.

More recent experimental evidence obtained from action spectra has shown that the sporangiophores probably possess more than one photoreceptor pigment molecule that participate in the photoprocess of phototropism (Galland and Lipson, 1984). However, this does not contradict the prior evidence that a flavin or flavoprotein is one of photoreceptor molecules that participates in the photoreceptor system of *Phycomyces* (Galland and Lipson, 1984; Presti and Galland, 1987).

The photoreceptor structure

This brings up the question: What is the photoreceptor structure and where is it located in the cell? The *Phycomyces* sporangiophore is nearly transparent and could act as a cylinder lens. The evidence for lens effects has been interpreted as indicating that the photoreceptor is located in or near the cell wall. Cohen and Delbrück (1959) showed that the primary effect of the light must be on some structure that moved relative to the cell wall. If so, a microscopic search should reveal a structure that moves in the growth-zone and responds to light. Observations with polarization microscopy revealed birefringent crystals (Figure 6.4a,b) that were aligned near the vacuole in the light growth-zone (Wolken, 1975). Further microscopic observations of the sporangiophores during growth revealed that these were octahedral crystals (Figure 6.4c). These crystals were isolated by ultracentrifugation of the sporangiophore (Figure 6.5) and their structure examined by microscopy and electron microscopy (Figures 6.5, 6.6, 6.7). It was found that the number of crystals per sporangiophore consistently increased with the sporangiophore growth Stages II-III, until early Stage IV, when their numbers decreased. In cultures grown in darkness, the number of octahedral crystals per fresh weight of sporangiophores was highest in Stage I and lowest in Stage IV, indicating that the number of crystals formed per sporangiophore was greatly dependent upon the light intensity. Continuous illumination reduced the number of crystals during all growth stages. In evaluating the number of crystals formed, it was found that there was a two-fold difference between dark- and light-grown *Phycomyces* (Ootaki and Wolken, 1973).

It is interesting to note (1) that more crystals were found in the wild-type than in the β-carotene-deficient mutants, which are only phototropic at high light intensity; (2) that the smallest number of crystals was found in the "night blind" mutants. Keeping in mind the possibility that β-carotene functions as a filter in the wavelength range of 400–500 nm; and (3) that since photodestruction of the crystals takes place at high light intensity, there should indeed be more crystals in the wild type, which contain a larger amount of β-carotene than the albino mutants.

The question remains whether the octahedral crystals found primarily in the light growth-zone are in fact the photoreceptors (Figure 6.5). If these crystals participate in the photoreceptor process, then their absorption spectra should mirror that for *Phycomyces* action spectra. The isolated crystals from Stage IVb absorption spectrum peaks are around 280 and 360 nm. However, in Stages I and

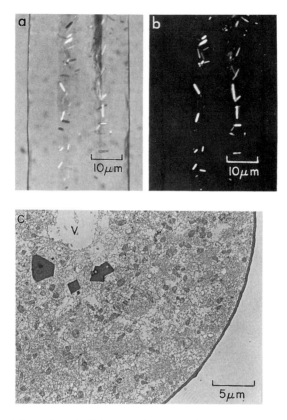

FIGURE 6.4   (a) Crystals in the light growth zone of *Phycomyces;* (b) same area as (a) in polarized light; (c) electron micrograph of a cross-section through light growth zone showing octahedral crystals (V = vacuole).

II-III, the isolated crystal absorption spectrum is around 275 to 285, 355, and around 465 nm. These are, in fact, the same absorption peaks found in the phototropic action spectrum and by microspectrophotometry of the sporangiophore light growth-zone (Figures 6.3, 6.6). The absorption in the ultraviolet around 280 nm indicates a protein, while the absorption around 355 nm and around 465 nm in the visible is indicative of an oxidized flavoprotein (Ootaki and Wolken, 1973). When these crystal are irradiated with light they are photosensitive and their spectrum resembles that of a flavin semiquinone or lumichrome (Figure 3.10b). These spectra and chemical analysis data suggest that, most likely, a flavoprotein is associated with these crystals. Chemical analysis of these crystals found that they were 95% protein and that the remainder consisted of lipids, flavins, and carotenes. Assuming that a flavoprotein and/or a carotenoid were adsorbed on the surfaces of the octahedral crystals, then 10^6 to 10^7 molecules

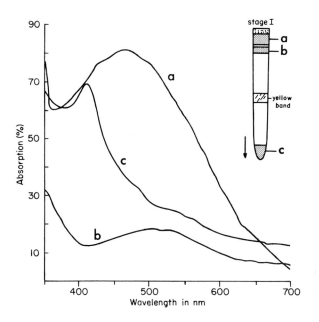

FIGURE 6.5   Absorption spectra of three crystal layers, a, b, c, in the centrifuged sporangiophore, Stage I; arrow indicates direction of centrifugal force. (obtained with a microspectrophotometer)

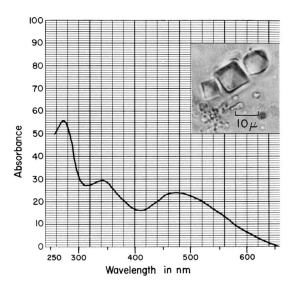

FIGURE 6.6   Absorption spectrum of an octahedral crystal (obtained with a microspectrophotometer).

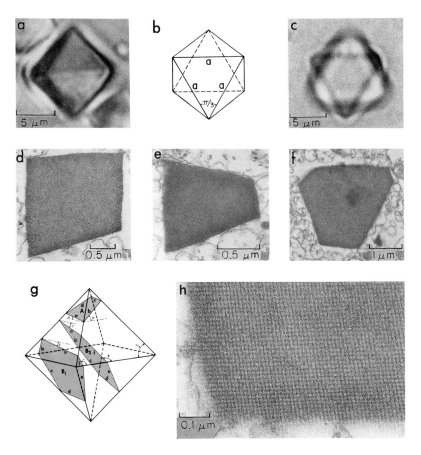

FIGURE 6.7   Octahedral crystal isolated from *Phycomyces*. Light microscopy (a) side view, (b) top view, showing equilateral triangular crystal faces as in (c). Electron micrographs (d–f) show the lattice structure obtained by various cuts through the octahedral crystal (g) and (h) enlargement of crystal structure.

could be accommodated. This is the right order of magnitude found for the number of photoreceptor molecules.

Since carotenes were detected in crystals, they were a likely place to search for retinal, the chromophore of visual pigment rhodopsin. To test whether retinal was simply absorbed on the crystal surfaces or actually part of the crystal itself, the isolated octahedral crystals from Stages II-III were fixed in 4% glutaraldehyde and scanned using a microspectrophotometer. The absorption spectrum showed a maximum around 520 nm, indicative of the visual pigment rhodopsin (Figure 6.8a). To identify whether retinal was found in these crystals, another test was made with freshly isolated crystals that were dried and reacted with the Carr-Price reagent ($SbCl_3$ in chloroform). The absorption spectrum of this reaction gave a peak at 664

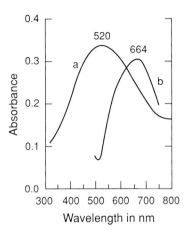

FIGURE 6.8   Isolated octahedral crystals from Stages II–III fixed in 4% glutaraldehyde; absorption spectrum was around 520 nm, indicating that of rhodopsin. (a) Crystals reacted with the Carr Price reagent (SbCl₃); absorption spectrum at 664 nm, indicating that of retinal (b).

nm indicating the presence of retinal (Figure 6.8b). These findings suggest that, in addition to a flavin, a retinal protein is probably one of the photoreceptor molecules and that other photoreceptor pigment molecules cannot be excluded when considering these photoprocesses of *Phycomyces* light growth and phototropism (Wolken, 1986).

PHYCOMYCES: A MODEL NEUROSENSORY CELL

> Phycomyces *is the most intelligent primitive eucaryote and as such capable of giving access to the problems in biology that will be central in the biology of the next decades.*
> —M. DELBRÜCK, 1976, *Light and Life*

Are the *Phycomyces'* sporangiophore behavioral responses to light analogous to those of a photo-neurosensory cell? If so, then upon light stimulation measurable electrical signals should be detected. The kinds of electrical signals that should be detected are: (1) an early receptor potential (ERP) which can be related to the photoreceptor pigment, (2) a receptor potential which is a positive and/or negative potential, and (3) a receptor potential, spikes, which is related to a nerve discharge.

When the Stage I sporangiophore was illuminated with a high light intensity source (60 Watt/sec), an early receptor response was recorded, indicating that it was related to the photoreceptor pigment. For all growth stages studied, there was a graded receptor potential, a positive wave whose amplitude correlated roughly with intensity, that was from 2 to 10 mV and from 2 to 10 sec in duration. At equal intensities the amplitude showed a larger response to a wavelength of 485 nm than to wavelengths of 420 or 385 nm. A receptor potential was also recorded in response to 500 nm in Stage I but not in Stage IV. One or more negative spikes occurred on an average of 2.5 min after exposure to the high intensity light source. These spikes reached amplitudes of 8 to 20 mV and durations of 0.5 sec. At 420 nm, spikes were observed but at a much lower intensity. These spikes, as with the early receptor potential response, occurred only in Stage I (Mogus and Wolken, 1974).

In response to electrical stimulation for Stages I, IVa, and IVb, small amplitude, biphasic potentials that rarely exceeded 10 mV and with a latency of about 1 min were observed. As the sporangiophore matures through Stages II to IVb, the complexity and type of the measurable electrical responses changes. Although the receptor potential is observed for all stages studied, the sporadic early receptor potential type response and the spike are not. This bears a relationship to the pigment system of *Phycomyces,* since orientation of the photoreceptor pigment molecule is related to the early receptor potential. The latency and time course of the receptor potential are slow when compared to that of animal visual photoreceptors (Tomita, 1970). The amplitude of the response is related to both the intensity and the wavelength. The electrical responses to light of 385, 420 and 485 nm mimic the absorption peaks in the phototropic action spectra for *Phycomyces* (Mogus and Wolken, 1974).

A latency of 2.5 min suggests a relationship with the transient growth response, which occurs within 3 min after exposure to light (Bergman et al., 1969). The absence of the spike response in Stage IV is also related to the pigment system of *Phycomyces.* It is possible that the photoreceptor pigment system in Stage IV may be in a different state from that in Stage I. The absence of both spike and early receptor potential type responses, as well as lack of response to 500 nm, suggests that different pigments mediate the photoprocess. These electrophysiological measurements show that when *Phycomyces* sporangiophores are under continuous illumination, their sensitivity to a given flash of light decreases. The loss of sensitivity as with photoreceptors in general is a function of the intensity of the incident light. This photoprocess in *Phycomyces* is reversible and it recovers to its original light-sensitive state. These experimental results indicate that *Phycomyces* sporangiophore photoreceptor system has a photosensitive pigment that is bleached by light and is resynthesized in the dark, a mechanism analogous to that found for the visual pigment rhodopsin in the retinal photoreceptors of the eye.

In summarizing these experimental findings, one can say that *Phycomyces* has evolved a highly sophisticated photosensory system, a sensor able to detect and locate the direction of light and to track a moving light beam. An example of its ability to sense the direction of light and store the direction of the light is found when *Phycomyces* spores are cultured in complete darkness until sporangiophores develop to Stage I, and, if given a brief flash of light through a pinhole, they will continue, in the absence of light, to grow in search of the light signal. By the time it reaches Stage IV, some 100 hours later, all the sporangiophores are observed to be bent directly toward the pinhole where the original light signal was received. In a sense the organism has a "memory" that records where the original light signal came from. There are other light responses of the *Phycomyces* sporangiophore photoreceptor system that are of interest, responses that can measure the intensity of light and select primarily blue light for phototropism. The absorption of light by the photoreceptor pigment system transduces the light energy to chemical energy for growth and to mechanical energy for phototropic movement. As a photo-neuro cell, measurable electrical signals are recorded during these photoprocesses. *Phycomyces* also possesses neurotransmitter molecules, such as acetylcholine and its enzyme acetyl cholinesterase that is found in concentrations of 10^{-8} moles/gram which is comparable to that found in neural cells (e.g., 10^{-6} moles/gram in brain cells). That *Phycomyces* should have incorporated such photo- and neurosensory mechanisms in its behavior is remarkable and provides us with a model cell to investigate more highly evolved photosensory systems.

PHOTOTAXIS

Since it is largely from the reaction of free unicellular organisms that our ideas of chemotaxis, phototaxis, and the like have been derived, it is important to study carefully the reactions of these creatures and to determine the laws which control them.
—HERBERT SPENCER JENNINGS, 1906, *Behavior of Lower Organisms*

In search of phototactic mechanisms, we can turn from phototropism to phototaxis. Phototaxis is observed in free-moving organisms (as in the swimming about of bacteria, algae, protozoa, and animals) when they move either toward or away from a light source. Organisms will orient themselves with respect to the direction of the light source by determining the intensity and wavelengths of the light in search of an optimum environment.

Anton van Leeuwenhoek in 1674, with his simple microscope, described the swimming pattern to light of algae and protozoa. "These animalcules had diverse colors, some being whitish and transparent; others with green and very glittering little scales; others again were green . . . and the motion of most of these animalcules in the water was so swift and so various, upwards, downwards, and round about, that "twas wonderful to see . . ." (in Dobell, 1958).

The swimming patterns of microorganisms have fascinated microscopists for a very long time, and such observations continue to intrigue biologists. For from these observations have come interesting behavioral relationships among the various organisms in response to light. The analysis of phototaxis has broad implications toward our understanding of photosensory mechanisms of more highly evolved organisms (Kreimer, 1994).

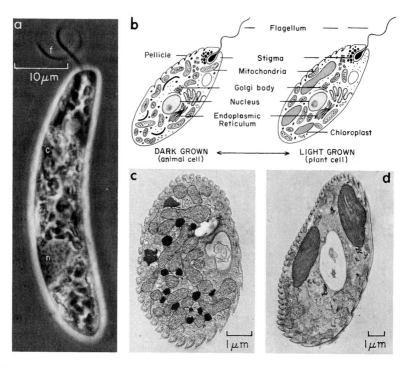

FIGURE 6.9   *Euglena gracilis* (a), cross-sections (c) of dark-grown and (d) light-grown (electron micrographs); structures are schematized in (b). (From Wolken, 1975, 1986.)

The protozoan algal flagellate *Euglena gracilis* is of special interest in studies of phototaxis. *Euglena* searches for light to develop chloroplasts and to utilize the energies of light for photosynthesis (Figure 6.9a,b,c). In the absence of light the organism degrades its chloroplasts and depends on chemosynthesis; it eats as an energy source, as do all animals for survival. *Euglena* is then truly both a plant and an animal cell. The metabolic processes are reversible depending on whether it lives in the presence or in the absence of light (Wolken, 1967, 1975, 1986).

The photoreceptor system for phototaxis

When *Euglena gracilis* is observed through a microscope, various kinds of motion can be distinguished: pulsation, sideways rotation, and forward swimming (Figure 6.10a). The sideways rotation and forward swimming are controlled by the whipping of the flagellum at the anterior end of the organism (Figure 6.10b,c). Light acts as a stimulus, and *Euglena* respond by swimming toward the light source— they are positively phototactic. At very high light intensity they are negatively phototactic and swim away from the high light source in search of an optimal lighting environment.

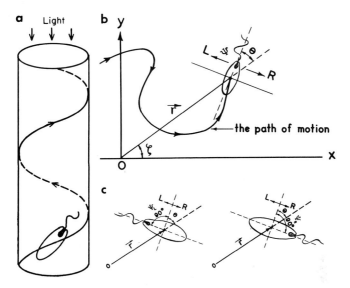

FIGURE 6.10   Swimming pattern of *Euglena gracilis* in response to light. (a,b) Paths of motion; (c) orientation and degree of left–right turning.

The *Euglena* photoreceptor system allows it to detect and respond to the direction of light. The photoreceptor system is referred to as the *stigma* and consists of the *eyespot,* the *paraflagellar body,* and the *flagellum* (Figure 6.11a). In some organisms, a lens-like structure is found near the eyespot. The photoreceptor structure is the paraflagellar body, which is attached to the flagellum and faces the eyespot. Electron microscopy of the paraflagellar body reveals that it is a membranous, crystalline structure with unit spacing of about 100 Å (Wolken, 1977). Optically diffracted reconstructed images from the electron micrograph of the paraflagellar body indicate that the structure is comprised of helically packed rods (Figure 6.11b,c,d). The eyespot cross-section is about 6 μm^2 and is an agglomeration of numerous orange-red pigmented globules that vary from 0.1 to 0.3 μm in diameter. These pigment globules are located just below a chamber with smooth walls that follows the rigid gullet from where the flagellum originates. In *Euglena,* the eyespot pigment globules are believed to act as a shading device and light filter for the photoreceptor.

Among algae *Chlamydomonas, Peridinium,* and *Volvox,* the eyespot pigment globules are found associated with the chloroplast membrane. The eyespot pigment globules in *Chlamydomonas* are arranged in layers. There are from two to about nine such globular layers (Figure 6.12). Foster and Smyth (1980), taking into account the spacing of these globules and their indices of refraction, suggested that the layered globules act as quarter wavelength interference reflectors for the organism.

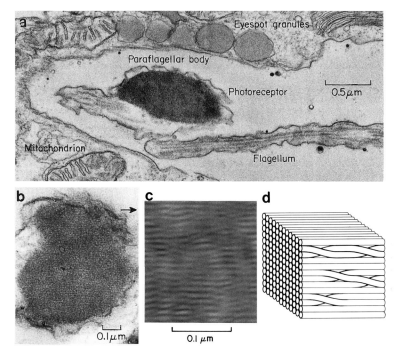

FIGURE 6.11 *Euglena gracilis.* (a) Eyespot area, eyespot granules, paraflagellar body, and flagellum. (b) Paraflagellar body and photoreceptor. (c) Optically diffracted electron micrograph image of paraflagellar body. (d) Schematic drawing of orientation of lamellae in the paraflagellar body. (From Wolken, 1977.)

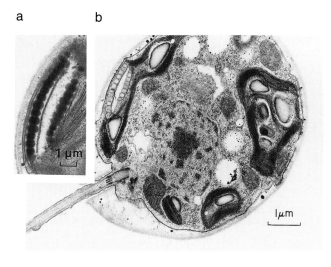

FIGURE 6.12 *Chlamydomonas,* cross-section (a) and enlargement of eyespot granules. (b) Electron micrograph. (Courtesy of Prof. J. Jarvik, Carnegie Mellon University, Pittsburgh, Pennsylvania.)

In search of the photoreceptor molecule

Euglena are phototactic; they swim toward light. The action spectra for the photo-behavioral responses to differing intensities and wavelengths of light should correspond to the absorption spectrum of the molecule or molecules responsible for such photobehavior in *Euglena*.

It was observed that the swimming velocity did not immediately change with light intensity, but it took 10 to 15 min for the organism to adapt to these light conditions before a regular pattern of phototactic response could be observed. A similar lag period of 10 to 15 min for a change in velocity following a change in illumination was also observed for *Chlamydomonas, Volvox,* and other algal cells. A possible interpretation of this lag time is that it depends upon a mechanism that is only indirectly affected by illumination.

The relationship between the swimming velocity of *Euglena* to varied light intensities is shown in Figure 6.13. It will be noted that the mean velocity or rate of swimming rises sharply from 0.11 mm/sec at 3 μW/cm^2 to a maximum rate of 0.16 mm/sec at 60 μW/cm^2, when it then starts to decrease slowly as the light intensity is raised above the saturation intensity (Figure 6.13). The significance of this value is that the number of light quanta at this particular intensity, 60 μW/cm^2, is sufficient to cover all the photoreceptor molecules that have a thermal energy equal to or greater than the minimal thermal energy, which is defined by the

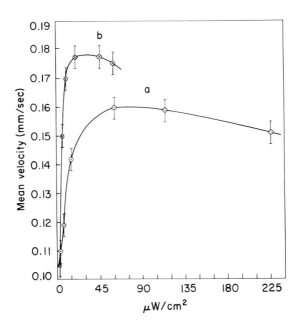

FIGURE 6.13 *Euglena gracilis.* The effect of different light intensities on the mean velocity or rate of swimming in (a) nonpolarized white light and in (b) polarized light.

illumination and the absolute temperature. Beyond this intensity, the absorption rate will remain constant without being disturbed by the extra number of quanta falling on the photoreceptor molecules, since the maximum absorption capacity has already been reached. The same effect is observed at various wavelengths, although the saturation intensity is different in each case.

In evaluating the various wavelengths of light to the relative rate of swimming, it was found that the mean velocity (mm/sec) plotted against light intensity gave straight lines for light intensities of less than 22 μW/cm^2. The action spectrum plotted for photokinesis, the rate of swimming (mean velocity in mm/sec against wavelength at 60 μW/cm^2 intensity) is shown in Figure 6.14. It will be observed that there is a major peak at 465 nm and another peak near 630 nm. This action spectrum is indicative of the absorption spectrum of the pigment involved in regulating the rate of swimming. The absorption peak around 465 nm is suggestive of a carotenoid or a flavin and the absorption peak at 630 nm is indicative of chlorophyll, or its precursor protochlorophyll. The action spectrum for *Euglena* photosynthesis and the action spectrum for O$_2$ evolution are found to be similar (Figure 6.15). This shows that *Euglena* searches for light so that it can efficiently use it for photosynthesis and implies that similar pigment molecules participate in both phototaxis and photosynthesis.

It will be noted (Figure 6.13) that there was a greater phototactic response to polarized light than in nonpolarized light and the action spectrum had absorption peaks around 465 and 500 nm with a rise beyond 600 nm. The polarized light effect probably indicates that there are two light-absorbing pigments, for example β-carotene and a flavin (Diehn, 1973). If there are two pigments, the resulting

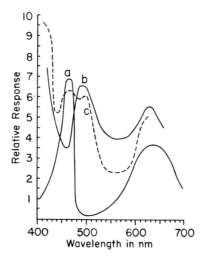

FIGURE 6.14   *Euglena gracilis.* Action spectra for (a) photokinesis, (b) phototaxis, (c) phototaxis in polarized light.

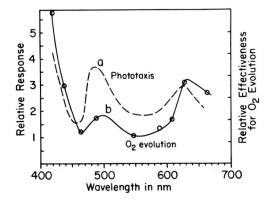

FIGURE 6.15   (a) *Euglena* phototactic action spectrum compared to (b) action spectrum for O_2 evolution in photosynthesis.

polarized light effect could be due to mutual energetic interference between them. On the other hand, the differences in *Euglena's* spectral sensitivity to polarized light may be due to the photoreceptor crystalline structure of the paraflagellar body that can function as a dichroic crystal photodetector (Foster and Smyth, 1980).

The photoreceptor pigment molecule

Englemann (1882) and later Mast (1911) have extensively studied *Euglena's* phototactic reactions to light. From their observations of *Euglena's* photobehavior, they speculated that these organisms possessed a photosensitive pigment molecule that resided in the eyespot area and had similarities to a visual pigment. If in fact *Euglena's* photoreceptor molecule for phototaxis were the visual pigment rhodopsin, then the action spectrum obtained for phototaxis would be similar to the absorption spectrum of a rhodopsin. However, the action spectrum obtained was that of a typical carotenoid. When their pigments were extracted from light- and dark-grown *Euglenas,* three main carotenoids were identified: β-carotene, lutein, and neoxanthin. Chromatographic analysis of the extracted *Euglena* carotenoids showed that, of these carotenoids, 80% were lutein or antheraxanthin, 11% were β-carotene, and 7% were neoaxanthin. There were also small amounts of cryptoxanthin, echinenone, and two ketocarotenoids, euglenanone and hydroxyechinenone. Any of these carotenoids could account for the absorption peaks found in the action spectra (Figure 6.14).

To specifically identify the photoreceptor molecule it is necessary to isolate the *Euglena* photoreceptor, the paraflagellar body, and to extract its pigments and chemically identify them. To isolate the paraflagellar body from the cell is difficult to do but can be circumvented by focusing on the photoreceptor through a microscope and obtaining the absorption spectrum with a microspectrophotometer. Absorption spectra of the *Euglena* eyespot area showed absorption peaks around 275,

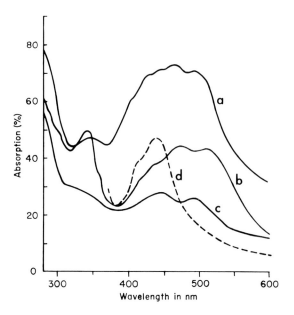

FIGURE 6.16   *Euglena gracilis.* Absorption spectra for (a) eyespot area, light-grown, (b) eyespot area of heat-bleached mutant, (c) eyespot area after dark-adaptation for 1 hour, and (d) eyespot area after 5 min white light.

350, 430, 465, and 495 nm (Figure 6.16a). In the heat-bleached (HB) mutant that lacks chloroplasts, and hence chlorophyll, the eyespot area spectra show similar absorption peaks. Spectra obtained closer to the paraflagellar body had absorption peaks at 440 and 490 nm and around 350 nm (Figure 6.16c). These spectra are similar to those found in *Phycomyces* crystals, which suggests that the photoreceptor molecule in *Euglena* is a flavin and that it participates in the photoprocess known as phototaxis. Further evidence for a flavin was obtained when light-grown *Euglena* were dark adapted for 1 hour, mounted on the cold stage (5°C) of the microspectrophotometer, and the eyespot area illuminated with strong white light for 1 to 5 min. The absorption peak (Figure 6.16d) around 490 nm bleaches, and there is an associated increase in absorption at 440 nm. This spectral shift on light bleaching is similar to that of a flavoprotein semiquinone, going from the reduced state to the oxidized state. If a flavin is the photoreceptor molecule, its identity and concentration should be more meaningful. The number of flavin molecules from light- and dark-grown *Euglena* was found to be of the order of 10^8 molecules per cell (Wolken, 1977). This is comparable to the number of chlorophyll molecules in the chloroplast and rhodopsin molecules in the retinal rods.

   The experimental evidence for phototropism and phototaxis of fungi, algae, and protozoa indicates that a flavin and/or a flavoprotein is the photoreceptor molecule (Figure 6.17). Flavins are also found to be associated with the electron transport

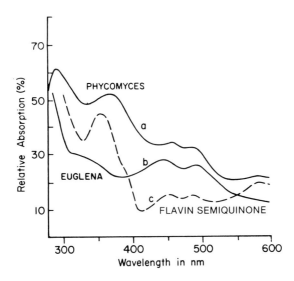

FIGURE 6.17   Absorption spectrum of light growth zone, photoreceptor area of *Phycomyces*, compared to (b) absorption spectrum near paraflagellar body of *Euglena* and to (c) absorption spectrum of a flavin semiquinone.

system of chloroplasts in algae, with light enhanced respiration in the alga *Chlorella* and with phototaxis of the dinoflagellate *Gyrodinium dorsum* and the insect *Drosophila* (Wolken, 1971, 1975). Nevertheless, these mechanisms for phototropism and phototaxis are very complex, and other photoreceptor pigment molecules probably participate in these photoprocesses.

Is *Euglena* a photo-neurosensory cell?

We can now consider *Euglena* as a primitive photosensory cell and examine more closely the organism's role as a photo-chemo-electro system. In *Euglena*, the photoreceptor, paraflagellar body, is associated with the flagellum, and together they serve as its photoreceptor-effector system. By means of the intensity and wavelength of light we can "communicate" with the organism to the extent that its rate of swimming and direction can be controlled. This suggests a photosensory cell or an analog of a photocell. *Euglena's* photoreceptor system may therefore be regarded as a servo or feedback mechanism which endeavors to maintain an optimal level of light. The photoreceptor and flagellum are linked so that light falling on the "eye" produces motion. This translation of an internal effect into a surface action produces similarities to those involved in nerve impulses in animal photoreceptor cells. The energy necessary to produce a perceptible mechanical response can be roughly calculated from the area of the photoreceptor eyespot region, the light intensity, and the effective wavelength. Using these data, the calculated energy was found to be 1.7×10^{-11} ergs/cm²/sec, or a quantum efficiency of 14%. This compares to the number of photons which can excite the eye.

The absorption spectrum of the eyespot also implies that the velocity of swimming is proportional to the number of light quanta absorbed by the photoreceptor and that the swimming motion is energetically controlled by light absorbed by the photoreceptor molecule. The shape of the intensity-dependence curve in Figure 6.13, showing a gradual rise with increasing intensity and the appearance of plateaus at higher intensity values, is very similar to the current-intensity curve of a photoconductive cell.

In the photoreceptors of higher animals, the photoexcitation triggering the optic nerve and, for the most part, the energy contained in a pulse is derived from chemical energy. Thus, the number of electronic charges involved in forming one such pulse is much larger than the minimum number of light quanta required to trigger the optic nerve. In the case of *Euglena,* however, such an amplification mechanism is not necessary. The minimum number of quanta required to excite the photoreceptor area is comparable to the power involved in the swimming motion. This means that one light quantum which is effectively absorbed by the photoreceptor can be associated with approximately one electronic charge formed at the base of the flagellum. At the saturation intensity of 60 μW/cm², the swimming velocity is about 0.18 mm/sec in a medium of viscosity of 0.987 centipoises. Using the cross-section of the eyespot, the intensity of 60 μW/cm² at wavelength 465 nm (which is equivalent to about 2×10^4 quanta/cm²/sec) and the average radius of the *Euglena* cell, the threshold potential was estimated to be of the order of 0.01 to 0.1 mV (Wolken, 1975). This is small when compared with the values found for animal nerves but is comparable to the early receptor potential (ERP) found in most excitatory tissue and is within the order of magnitude for certain insect eyes (Naka, 1960).

There are other considerations in regard to the mechanisms of phototaxis. How is the light energy that is absorbed by the photoreceptor transduced to chemical energy that triggers the flagellar motion? The flagellum consists of a number of elementary filaments, *axonemata,* that are embedded in a matrix and covered by a membrane (Figure 6.11). The flagellum of *Euglena* is of the order of 30 μm or more in length and has numerous junctions along its entire length. In cross-section the flagellum is from 0.25 to 0.40 μm in diameter. Its internal structure consists of an axoneme containing the (9 + 2) pattern of filaments running the length of the flagellum. The two central filaments are typical microtubules, while the nine peripheral filaments are microtubular doublets. This structural arrangement of the flagellum is found with all motile cells. It is found, for example, from bacteria to vertebrates, including the sperm tails of humans and the connecting filament between the outer segment and the inner segment in vertebrate retinal rods (Figure 9.2c). The chemistry of the algal flagella is composed like that in muscle of a myosin-like contractile protein (Lewin, 1962; Witman et al., 1972a,b). When large quantities of *Euglena* flagella are isolated and placed in 10^{-3} M solutions of

ATP, they display vigorous beating (Wolken, 1967, 1975). So we see that the flagellar motion is driven by a photo-chemo-mechano-mechanism that, in the process, converts the chemical energy of ATP into movement—like that of a muscle fiber.

Euglena cell behavior has similarities to a photo-neurosensory cell in its mechanisms of excitation. As a neurosensory cell it should possess neurotransmitter molecules. In fact, *Euglena* has of the order of 3.85×10^2 acetylcholine molecules (Wolken, 1971). Such a neurotransmitter molecule is specifically associated with the chemistry of nerve excitation in the nervous tissue. Neurotransmitter molecules are found in protozoans and other relatively primitive organisms (Lentz, 1968). It seems reasonable that the basic secretory capacity of these pre-nervous cells was modified during evolution for the coupling of the excitatory and conductive properties to allow transmission to occur upon light absorption or by chemical and electrical stimulation. Therefore, we can look upon *Euglena* as a primitive "retinal" cell (Figure 9.12).

CONCLUDING REMARKS

Photoreceptors for light-searching arose among bacteria, fungi, algae, and protozoa early in the history of life. With a photoreceptor system these organisms are able to detect the direction of light, measure its intensity, select the wavelengths of light, and move toward or away from the light source in search of an optimal environment—necessities for survival.

An example chosen from among these unicellular organisms is the phototropic fungus *Phycomyces*. Its photoreceptors are crystals that reside in the light growth-zone of the cell sporangiophore. The photoreceptor molecule is a flavoprotein, but other pigments, such as carotenoids, cytochromes, and a retinal, most likely participate in the photoprocesses of light growth and phototropism. In *Euglena*, a protozoan algal flagellate, the photoreceptor system for phototaxis consists of eyespots, a photoreceptor paraflagellar crystalline body, and a flagellum (Wolken, 1977). The photoreceptor molecule is a flavoprotein, as in *Phycomyces*, but retinal and other pigment molecules, as in *Phycomyces*, cannot be excluded from the photoprocesses. In the protozoan *Paramecium busario*, a retinal protein was identified as the photoreceptor molecule (Tokioka et al., 1991). The paraflagellar body was isolated from *Euglena*, and in addition to flavins, pterines were found (Brodhum and Häder, 1990). Pterines have also been found in association with flavins in the *Phycomyces* sporangiophore (Kiewisch and Fukshansky, 1991). The photoreceptor structure in these organisms is a "crystalline" body whose lattice spacing provides for an ordered molecular structure for the orientation of the pigment molecules to maximize light reception for transduction to chemical energy.

In the algal flagellate *Chlamydomonas*, the photoreceptor is composed of the pigmented eyespot globules that are attached to the chloroplast (Figure 6.12). The photoreceptor pigment in the eyespot globules was surprisingly found to be a rhodopsin, the visual pigment of animal eyes. *Chlamydomonas* then has its photoreceptor system—the eyespot for light searching and the chloroplast for photosynthesis. In the halophilic bacterium, *Halobacterium halobium,* the photoreceptor pigment resides in its cell's purple membrane, and the pigment molecule is also a rhodopsin. The discovery of rhodopsin as a photoreceptor pigment in these organisms confirms the earlier speculations of Englemann (1882) and Mast (1911) and indicates that the visual pigment rhodopsin was synthesized for light reception well before it became incorporated into the photosensory cells and the retina of animal eyes for visual excitation.

Much more research is needed to decipher the mechanisms of phototactic behavior from a molecular, structural, photochemical, and genetic standpoint in order to discover the details of light reception and signal processing in living cells. Therefore, investigations of phototropism and phototaxis, in these unicelluler organisms, has broad implications toward elucidating photosensory mechanisms of more highly evolved animals.

NATURE'S WAYS
OF MAKING IMAGES
FROM LIGHT

CHAPTER SEVEN

Emergence of an Imaging Eye

> *All that is needed as the starting point for the development of eyes is the existence of light-sensitive cells.*
>
> —ERNST MAYR, 1982, *The Growth of Biological Thought*

> *The principal means by which most animals are made aware of their surroundings . . . is the reflection or emission of light toward them by external objects and the reflection of this light by special organs—photoreceptors. The more complicated of these photoreceptors are called eyes.*
>
> —GORDON L. WALLS, 1942, *The Vertebrate Eye*

How did the eye evolve? In examining the evolutionary phylogenetic scheme of animals from protozoa to vertebrates (Figure 7.1), where among these phyla did an eye evolve? In search for the evolutionary development of an eye Walls (1942), Duke-Elder (1958), Willmer (1960), and Eakin (1965) have reviewed its structural development along phylogenetic lines. The evolutionary biochemical synthesis of visual pigment were investigated by Wald (1952) and Crescitelli (1972). These classical studies have contributed significantly to the understanding of how the eye and visual system have evolved. More recent experimental studies have taken another approach by using methods of molecular genetics to identify the genes that determine the synthesis of the visual proteins and visual pigments (Nathans, 1986 a, b). These experiments regarding the synthesis of visual pigments and the development of visual photoreceptors are reviewed by Goldsmith (1990). The development of imaging eyes and their optical systems is reviewed by Land (1989) and Nilsson (1989 a, b). These researchers have renewed an interest in the evolutionary development of the eye and visual system that is being pursued by biologists, biochemists, biophysicists, and bioengineers.

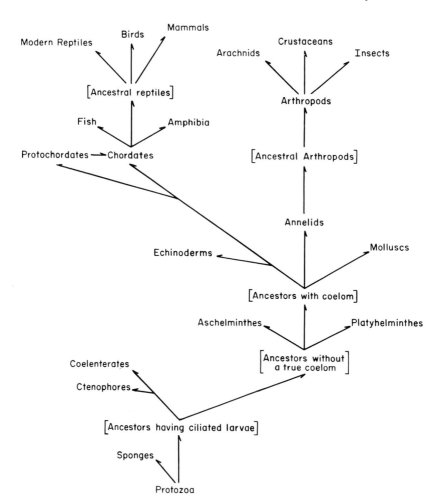

FIGURE 7.1   Phylogenetic evolutionary development of animals (after Wolken, 1986).

Here, I would like to go back and attempt to develop a logical sequence of events that led from the photoreceptors in unicellular organisms to more highly evolved multicellular organisms, photosensory cells, which later became incorporated into the retina, were covered by a lens, and emerged as an imaging eye. To trace out this evolutionary development, I began with the photoreceptor structures of algae, bacteria, fungi, and protozoa that were discussed in Chapter 6. To recollect, in the bacterium *Halobacterium halobium,* the photoreceptor structure resides in the pigmented "purple" cell membrane; in algae of the *Chlamydomonas* species, the photoreceptor system is comprised of pigmented eyespots surrounded

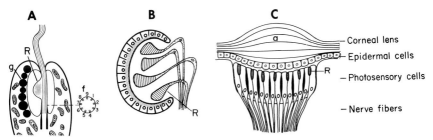

FIGURE 7.2   Photoreceptor systems: (a) eyespot flagelium, (b) *ocellus*, photosensory structure of flatworms, (c) structure of an *ocellus*.

by a membrane that is attached to the chloroplast, the photoreceptor for photosynthesis. The photoreceptor pigment molecule for both of these organisms is a rhodopsin, the visual pigment in the retina of all animal eyes.

In the algal flagellate *Euglena gracilis,* a more structured photoreceptor system is found, whose photoreceptor consists of the paraflagellar body, a highly ordered crystalline structure, the pigmented eyespots, and a flagellum (Figure 7.2a; see also Figure 7.10a). Some euglenoids also have lens-like structures (Fauré-Frémiét, 1958). Further elaboration of the photoreceptor system is found in the marine protozoan *Erythropsis,* in which the photoreceptor is located in a cup of pigmented globules and is covered by a large transparent spherical lens (Figure 7.3). Such an "eye" structure found among marine protozoa was referred to as an organ "ocu-

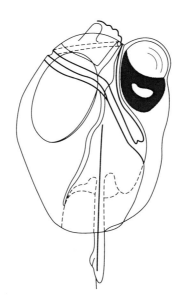

FIGURE 7.3   The "eye-oculare" ocellus of the protozoan dinoflagellate *Erythropsis* (after Kofoid and Swezy, 1921).

laire", an *ocellus,* or a simple eye, all of which are described in Kofoid and Swezy (1921).

These findings suggest that a photosensitive pigment, a rhodopsin, became an integral part of the cell membrane that formed into a multilayered photoreceptor structure within the cell and became a photosensory cell. Such photosensory cells later became structured into a retina for the eye.

The first "eyes" were probably photosensory cells over the animal's outer skin. An evolutionary step must have occurred when the photosensory cells invaginated into a cup in the animal's body, a hole in the skin over the cup acted as a lens and later became covered by an actual lens to focus the light on to the retina. Such an eye was connected by a nerve to carry visual information to the brain. As a result, animals with eyes were able to gain additional information about their world, thus greatly aiding their adaptation to life on earth.

Haldane (1966) stated: "There are only four possible types of eyes, if we define an eye as an organ in which light from one direction stimulates one nerve fiber. There is a bundle of tubes pointing in different directions and three types analogous to three well-known instruments, the pin-hole camera, the ordinary camera with a lens, and the reflecting telescope. A straight forward series of small steps leads through the pin-hole type to that with a lens, and it is quite easy to understand how this could have been evolved several times."

PINHOLE EYE

The simplest optical system for imaging is the "pinhole eye." A small hole in the wall of an opaque chamber allows the passage of very narrow beams of light from each point on an object which will form an inverted image on the opposite wall of the "eye" chamber (Figure 7.4a). As an image-forming eye, it is not very efficient, for only a small fraction of the light from an object can get to the photoreceptors. If the hole, or aperture, is made larger to increase the amount of light, image definition is lost; if it is made smaller to improve the resolution, diffraction effects

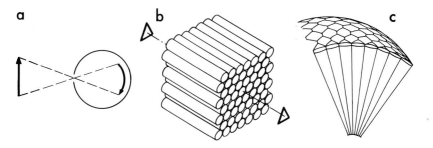

FIGURE 7.4   Simple imaging devices that evolved: (a) pin-hole, (b) parallel tubes, as in compound eyes (c).

become a problem. Nevertheless, the pinhole eye has the advantage of simplicity; no focusing is required, and the size of the image is inversely proportional to the distance of the object.

An example of a pinhole imaging eye is found in the cephalopod mollusc *Nautilis,* found in deep tropical waters and in areas of the Pacific Ocean. The Nautilis eye is remarkable and is highly adapted to its environment. The eye is unusually large and is formed by invagination of the photosensory cells in the skin into a hole. Structurally, the eye resembles a vertebrate eye, except that it does not have a lens, and the interior of the eye, by way of an open pupil, is directly in communication with the surrounding ocean water (Muntz 1987; Muntz and Raj, 1984; Muntz and Wentworth, 1987). The retinal photoreceptors are rhabdom-like structures similar to the retinal photoreceptors of arthropod compound eyes.

A simple eye must have evolved when the photoreceptor cells of an animal's skin (cuticle) invaginated to form a pigmented cup with a small opening (see Figure 7.2b,c). This provided the animal with photoreceptor cells exposed to the environment for light reception. This type of simple eye is found among flat-worms, coelenterates, annelid polychaetes, molluscs, echinoderms, insects, and protochordates. From among these animal phyla a few examples were selected to indicate the structure and diversity of their eyes.

In the phylum Platyhelminthes, the flatworms, the common planarian possesses two such simple eyes. Each eye is in a pigmented cup comprised of photosensory cells. The pigment cells function to shade the photosensory cells from light in all but one direction. This enables the planarian to respond differentially to the direc-tion of light; that is, to turn away from it—planarians are negatively phototactic. The animal's tendency to avoid light is controlled by the balance of nervous impulses from the photosensory cells of the eye. In *Planaria musculata,* the pigment cup is about 45 μm wide and 25 μm deep with an aperture 30 μm wide. The photoreceptor cell bodies are located outside the eye and joined to the nerve terminals via dendrites passing through the aperture. In other flatworms the photo-receptors point toward the light, with the cell bodies and nerve axons penetrating the pigment cup, which is more typical of simple eyes. Their photoreceptors are differentiated structures of the photosensory cells. They are generally about 5 μm in diameter and about 35 μm in length. Their microstructure is that of microvilli, typical of the rhabdomere photoreceptor structure of insect and crustacean com-pound eyes (Wolken, 1971). Similarly structured simple eyes were described for planaria *Dugesia lugubris* and *Dendrocoelum lacteum* (Röhlich and Török, 1961; Röhlich, 1966) and for the marine planarian *Convoluta roscoffensis,* found in Roscoff off the coast of Brittany in France (Keeble, 1910; Wolken, 1971, 1975).

CAMERA TYPE EYE

The development of the imaging eye required an optical system; a lens to focus the light and photosensitive receptor cells connected to a nerve to carry the imaged

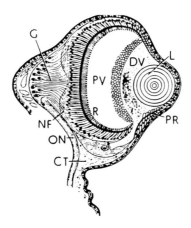

FIGURE 7.5   The polychaete worm *Vanadis*. CT, connective tissue; DV, distal vitreous; G, ganglion cells; L, lens; ON, optic nerve; NF, optic nerve fibers; PR, proximal retina; PV, proximal vitreous; R, retina showing rods separated from visual cell bodies by a dense line of pigment (after Hesse, 1899).

information to the brain. In the phylum *Annelida,* there are a variety of animals with eyes that have distinct lenses. Land (1980) has reviewed the structure and optics of annelid eyes. Among the annelids are the Alciopids, a family of marine, carnivorous, pelagic polychaetes. They are transparent and relatively large, growing up to 20 cm in length. Their eye structure was described by Greeff (1877), and Hesse (1899) specifically studied the eyes of *Alciopa cantrainii* and *Vanadis formosa.* Their eyes are about 0.5 mm in diameter and more highly developed than other polychaetes. Since they capture prey, the question arises: How good are their eyes? More recently, Hermans and Eakin (1974) studied the structure of the eyes of *Vanadis tangensis.* The eye of *Vanadis* is about 1 mm in diameter, including photosensory cells and a distinct lens. The photoreceptors are about 80 μm long and 6 μm in diameter, and formed of microvilli (Figure 7.5). Wald and Rayport (1977) determined the spectral sensitivity from the electroretinogram (ERG) of eyes of *Torrea candida,* a surface worm, and *Vanadis,* found in the deep sea, and concluded that they have image-resolving eyes. Therefore, highly developed imaging eyes evolved in annelid polychaetes.

In the family *Sylladae,* the polychaete *Odontosyllis enopla,* "the fireworm of Bermuda," the behavior of the visual system is of interest. *Odontosyllis* possesses a lunar periodicity; an hour after sunset at the time of the full moon, the female bioluminesces with a flashing blue-green light, which brings the male, which also bioluminesces to the female during its mating period. During mating the effect is a beautiful dance of light that lasts about thirty minutes each night and lights up the coves of Bermuda for about three days.

Odontosyllis is about 25 mm long and has four eyes arranged so that two pairs are located adjacent to each other on the dorsal surface of the head (Figure 7.6a–c). The eyes are on protruding lobes that can move. One eye is designed to look up from the water and the other to look forward. The eyes of the males are larger than the eyes of the females. The front of the eyes appears to be completely covered by the worm's cuticle, and each eye has a rounded exterior with a relatively small opening and a lens located behind the opening which lies in a cavity formed by the pigment cup. The lens is a spheroidal body composed of cells and is relatively large compared to the size of the eye. Under the lens are photoreceptor cells. The photoreceptor cells are long, structured membrane processes, or microvilli, like those of the mollusc cephalopod eye photoreceptors (Figure 7.7c–f). Also within the lens area and closely associated with the photoreceptors are long tubular rods, about 55 nm in diameter, arranged in linear arrays, suggesting that they may be fiber optic bundles or light guides (Figure 7.7g–h). A fiber optic system would function to detect the direction of bioluminescent light flashes and would maximize the light collecting of the eye (Wolken and Florida, 1984). For the worms, successful mating depends upon the detection and location of the bioluminescent light.

The spectral sensitivity as measured by the ERG is around 510 to 520 nm (Figure 7.8b), which coincides with the bioluminescent emission peak that of luciferin around 507 to 516 (Figure 7.8a) indicating that their eyes detect the bioluminescent light (Wilkens and Wolken, 1981). To determine whether their eyes had the visual pigment rhodopsin, the eye pigments were extracted, and a photosensitive pigment was isolated that had absorption peaks around 330 nm, 430 nm, and 500 nm. The absorption peak around 500 nm corresponded to the spectral sensitivity peak and was indicative of the visual pigment rhodopsin. Therefore, among the annelid polychaetes, an evolutionary development occurred from a simple eye to a camera-type imaging eye (Wolken, 1986).

Among the coelenterates are rotifers, phylum Achelminthes. Rotifers are similar to flatworms and are found in freshwater lakes and ponds. They are microscopic, under 1 mm long. Their eye structures are described as ocelli, and in many rotifers the ocelli are paired. The ocelli have photosensory cells and a lens that is associated with a red-pigmented eyespot (Clement et al., 1983; Duke-Elder, 1958).

An outstanding example of a camera-type imaging eye is found in the cubomedusan jellyfish of tropical waters. This jellyfish has as many as 24 eyes. Each eye has an epidermal cornea, a spheroidal lens, and retinal photoreceptor layers. The eye structure is similar to that described for alciopids. Therefore, coelenterates can now be added to the list including annelids, molluscs, and vertebrates with highly developed eyes (Pearse and Pearse, 1978).

Insects also possess from one to three or more ocelli. The median or dorsal ocelli of many insects consist of a layer of photoreceptor cells, a synaptic zone in which the axons of the photoreceptor cells come into contact with dendrites of the ocular nerve fibers, and an ocular nerve that leads from the eye to the brain (Chapell and Dowling, 1972). The role of the ocelli is not to resolve images but to

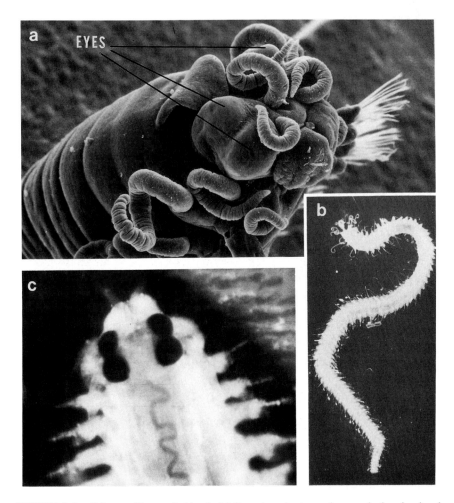

FIGURE 7.6 *Odontosyllis enopla* Verril. (a) Scanning electron micrograph showing head and eyes; (b) live female (2 mm long, 1.5 mm in diameter); (b1) model of the eye (magnification, 23×). (Continued in Figure 7.7.)

detect rapid changes in light intensity, an aid in escaping predators by sensing their shadows (Pollock and Benzer, 1988).

PROTOCHORDATES

Protochordates comprise three subphyla: Hemichordates, Urochordates, and Cephalochordates. These have traditionally been considered subphyla of the phylum Chordata, of which the fourth subphylum includes vertebrates. The structure

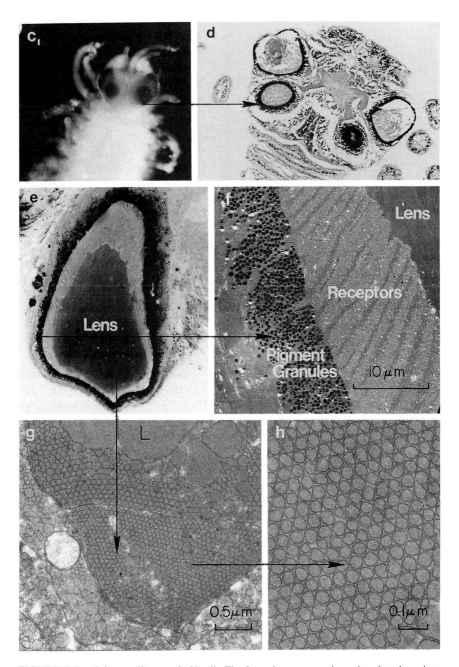

FIGURE 7.7 *Odontosyllis enopla* Verril. The lens, in cross-section, showing the micro-tubules in linear array. Electron micrograph. (c) Flattened area of the head showing the two eyes on each side of the head; (d) cross-section through eyes; (e) enlargement of eye; (f) section through photoreceptor area, lens photoreceptors, and pigment granules; (g) enlargement of lens (L) area; (h) microtubule in lens area.

97

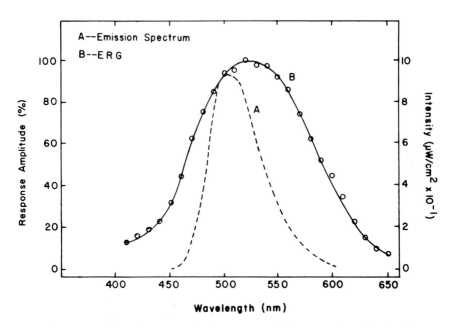

FIGURE 7.8   The bioluminescent emission spectrum (A) compared to (B) the electro-retinogram (ERG) spectral peak of the eye of *Odontosyllis enopla.*

of their photoreceptors may be a link to the development of the retinal photorecep-tors of vertebrate eyes. Early theories regarding the evolution of vertebrate retinal photoreceptor cells indicated that they may have arisen from a cephalochordate, *Amphioxus,* the lancet. *Amphioxus* do not have eyes, but they do have two kinds of pigment cells; one kind comprises the large pigmented eyespot at the anterior tip of the nerve cord. These pigmented cells in the central nervous system were believed to be the phototactic photoreceptors that controlled the direction in which the animal swam. The other kind of pigment cells comprise a group of ependymal cells connected with nerve fibers, the infundibular organ that is shielded by the pigmented eyespot. It was these ependymal cells that were thought to be homolo-gous with the visual photoreceptors of vertebrate eyes, but there is no experimental evidence to support this hypothesis (Walls, 1942; Willmer, 1960).

Clues, though, can be found in the tunicate ascidians, or sea squirts, of the subphylum Urochordata. They are known to be related to vertebrates because they have motile, tadpole-like larvae with definite chordate characteristics. The sea squirt, *Ciona intestinalis,* is of special interest in our discussion of the evolution of visual photoreceptors. *Ciona* is situated in the transition zone between inverte-brates and vertebrates and possesses structural features common to both. The *Ciona intestinalis* tadpole has a relatively simple ocellus that serves as its "eye" to detect the direction of light, thereby permitting the tadpole to orient itself and move toward or away from a light source. Dilly (1964, 1969) investigated the

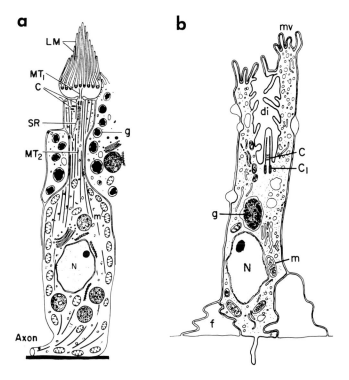

FIGURE 7.9   Schematic of (a) *Ciona intestinalis* larva; LM, sensory lamellae derived from cilium; MT_1, microtubule (in axoneme of cilium); C, centrioles; SR, rootlet of cilium; MT_2, microtubules in cytoplasm (according to Eakin, 1963). (b) *Ciona intestinalis*. Adult schematic of photoreceptor cell. C, axoneme of cilium; N, nucleus; g, pigment body. (From Dilly and Wolken, 1973).

structure of the ocellus and found it to be made up of about ten cells, comprised of four to nine retinula cells containing pigment granules and a lens cell. The retinula cell photoreceptor, according to Dilly, is structured of microvilli and is connected by a cilium (Figure 7.9); Dilly and Wolken (1973) also noted that there was a structural relationship to the vertebrate retinal rod cell.

Eakin and Kuda (1971) then reexamined the receptor structure of the ocellus of two species of ascidian tadpoles, *Ciona intestinalis* and *Distaplia occidentalis*. They found that the ocellus of *Distaplia occidentalis* is composed of one pigmented, cup-shaped cell with fifteen to twenty photosensory cells and three lens cells. Each photosensory cell possessed an outer segment consisting of many lamellae, microvilli, and an inner segment bearing the basal body of the modified cilium (axonemes of 9 + 0 doublets of microtubules). They thereby established the ciliary nature of the photoreceptor that would relate it to that of the vertebrate visual retinal rod outer segments. The relative evolutionary significance of the

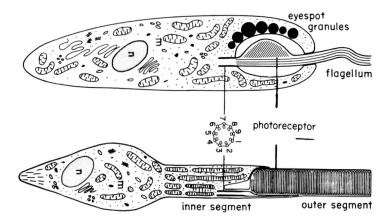

FIGURE 7.10   A structural comparison between the photoreceptor structure of the *Euglena* cell with that of the vertebrate retinal rod cell. (From Wolken, 1986).

adult and larval ocellus of *Ciona* suggests that the photoreceptor structure of the larva is on the main evolutionary pathway to the vertebrate retinal cells. Eakin (1965) had proposed that there was an evolutionary structural relationship between two types of photoreceptor structures and classified them depending on their origin as either *ciliary* or *rhabdomeric*. Those where the photoreceptors evolved from the cell membrane are of the invertebrate rhabdomeric type photoreceptors and those that arose from the cilium (or flagellum) are of the vertebrate type photoreceptors. Evidence for this can be found in comparing the photoreceptor structure of Euglena to that of the vertebrate retinal rod cell (Figure 7.10), indicating that both photoreceptors may have arisen from the flagellum (or cilium).

Therefore pinhole and simple eyes, ocelli and imaging eyes evolved independently among diverse lower animals, including flatworms, annelid polychaetes, coelenterates, echinoderms, insects, and cephalopod molluscs.

CHAPTER EIGHT

Visual Pigments

In vision, light plays an entirely different role; not to do chemical work, but—through the excitation of the visual pigment to trigger a nervous excitation.

—GEORGE WALD, 1973, *Biochemistry and Physiology of Visual Pigments*

VISUAL PIGMENTS

The chemical structures of the carotenoids, chlorophylls, phytochromes, flavins, and other accessory pigment molecules were previously described in Chapter 3. In discussing non-imaging system phototropism and phototaxis, it was indicated that a retinal-protein, the visual pigment rhodopsin, was one of the photoreceptor pigment molecules.

To initiate the visual process of "seeing" requires a visual photoreceptor pigment molecule. This brings us to inquire: What are the visual pigments, how are they chemically structured for photoreception in the retina of the eye, and how do they function for visual excitation?

A long history of discoveries led from the description of the retina by van Leeuwenhoek in 1674 to the isolation and chemical identification of the visual pigment. Two centuries after the initial description of the retina, Krohn (1842) and Müller (1851) observed that the frog and squid retinas were pigmented "red." Schultze (1866) fixed the retina in osmium tetroxide (OsO_4), a cellular fixative still in use for electron microscopy, and was one of the first to describe the structure of the retinal photoreceptors. This was followed by the observations of Böll (1876) and Kühne (1878) that the retinal rods were pigmented "reddish purple", when exposed to light bleached to yellow, and when returned to darkness, the reddish color was regenerated. They recognized that the bleaching and regener-

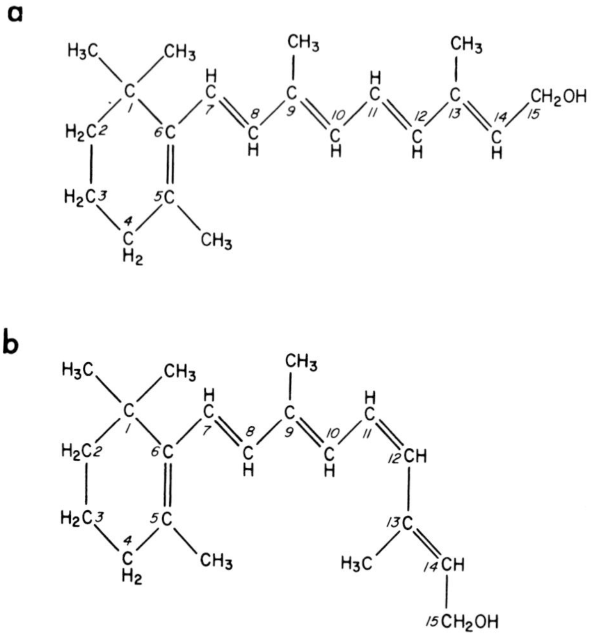

FIGURE 8.1   Chemical structure vitamin A: (a) all-*trans* and (b) 11-*cis*.

ation of the pigment was related to the visual process of seeing. Soon after, Kühne (1878) extracted the pigment from the retinal rods and named the retinal pigment *visual purple*.

The chemical identity of visual purple had to await the discoveries by George Wald (1933, 1935), who found that the biological activity of vitamin A was related to the synthesis of visual pigments and that vitamin A aldehyde was the chromophore *retinal* of the visual pigment rhodopsin (Figures 8.1, 8.2). For this discovery and for numerous investigations of visual pigments Wald was awarded the Nobel Prize in physiology and medicine in 1967.

Vitamin A_1 (retinol), vitamin A_2 (dehydroretinol), vitamin A acid (retinoic acid), and other natural derivatives of vitamin A are retinoids, so named because they were originally discovered in the retinas of animal eyes. These molecules exert a profound effect on growth of cell differentiation and development and are necessary for sustaining all animal life. Vitamin A in humans is stored in the liver and is carried by the bloodstream to the eye. The Egyptians recognized the potency of the juices of the liver in ancient times and used it to treat eye disorders and night blindness. We now know that when the stores of vitamin A in the liver and bloodstream have been exhausted, the first symptom of vitamin A deficiency in humans is the rise of the visual threshold to that of night blindness.

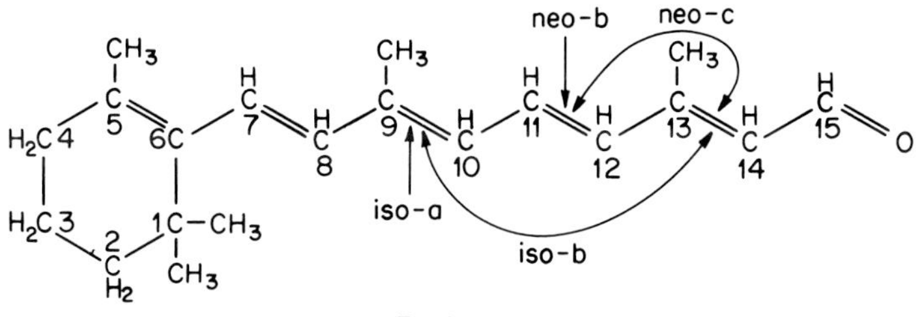

Retinal₁

FIGURE 8.2   Retinals, indicating different geometric isomers.

TABLE 8.1   Geometric Isomers of
Retinal

Isomerization around bonds	Nomenclature
9–10	9–*cis*
11–12	11–*cis*
13–14	13–*cis*
9–10, 13–14	9, 13–di–*cis*
11–12, 13–14	11, 13–di–*cis*

We have already noted that animals cannot synthesize C_{40}-carotenoids and need to obtain β-carotene by ingesting plants. The ingested β-carotene is metabolized by animals to a degraded derivative, a C_{20}-carotenoid molecule, of vitamin A (Figure 8.1). This indicated that there was a relationship between the ingested β-carotene to vitamin A and the visual sensitivity of the eye. In the synthesis of the conversion of β-carotene to vitamin A_1, retinal₁ is an intermediate product. The *in vivo* synthesis from β-carotene → vitamin A → retinal is of special interest from a biochemical and evolutionary standpoint.

To learn about the tissue function of vitamin A, vitamin A acid (retinoic acid) was substituted for vitamin A and was found to maintain growth in the rat, with a biopotency equivalent to vitamin A (van Dorp and Arens, 1947). However, no matter how large the amount of vitamin A acid fed to the rat no vitamin A was deposited in the liver. The rat was unable to reduce vitamin A acid to vitamin A, the form in which vitamin A is stored. This finding led Moore (1953) to suggest that vitamin A acid fulfills the tissue function of vitamin A, but is not able to serve as the precursor for the visual pigments, which require vitamin A for their synthesis.

Let us briefly review the synthesis of retinal, the chromophore of the visual pigment rhodopsin, from its precursor molecule, vitamin A. In the metabolism of vitamin A in animals, Morton (1944) and Morton and Goodwin (1944) found that retinal was an intermediate product in the metabolic process. They and Hawkins and Hunter (1944) found that retinal$_1$ was synthesized from vitamin A$_1$. Then Hunter and Williams (1945) demonstrated that by oxidation of β-carotene to Vitamin A, retinal$_1$ was an intermediate product. Then it was demonstrated that retinal$_1$ was rapidly converted to vitamin A$_1$ when it was administered orally, subcutaneously, or intraperitoneally (Glover et al., 1948). The conversion of retinal$_1$ to vitamin A$_1$ was a reduction which occurred in the gut and subcutaneous tissues. Then Hunter and Williams (1945) demonstrated that retinal$_1$ could be obtained by the oxidation of β-carotene.

Retinal$_1$ in solution can be converted back to vitamin A$_1$ by adding a reducing agent, thus shifting the retinal absorption peak of around 370 nm to 325 nm, that of vitamin A. A plausible explanation for the displacement of the absorption maximum of vitamin A$_1$ from around 325 nm to around 370 nm, that of retinal$_1$, is explained by an increase in the number of conjugated bonds from 5 to 6. The replacement of the terminal alcohol —CH$_2$OH group of vitamin A$_1$ by the aldehyde —CHO provides the sixth conjugated bond in retinal (Figure 8.2).

Vitamin A$_1$ can be identified by its absorption spectrum, for its maximum peak is around 325 nm (in hexane or in ethanol), and when reacted with the Carr-Price reagent (SbCl$_3$ in chloroform), the absorption peak is at 621 nm. Retinal$_1$ can also be identified by its absorption spectrum. In hexane the absorption peak is at 368 nm, in ethanol it is at 383 nm, and when reacted with the Carr-Price reagent, the absorption peak is at 664 nm (Hubbard et al., 1971). Vitamin A$_2$(-dehydroretinal) is a chemical isomer of vitamin A$_1$ and differs from vitamin A$_1$ in possessing an added double bond in the ring, at positions 3 and 4 (Figure 8.1). Vitamin A$_2$ also reacts with the Carr-Price reagent to give a maximum absorption peak around 693 nm. Maximum absorption peak for retinal$_2$ is around 385 nm in hexane, 388 nm in petroleum ether, and 401 nm in ethanol. Retinal$_2$ (3-dehydroretinal) also reacts with the Carr-Price reagent, with an absorption maximum around 705 nm. Therefore, from their chemical structures and absorption spectra, vitamin A$_1$ and vitamin A$_2$, and hence retinal$_1$ and retinal$_2$, can be distinguished from one another.

Vitamin A$_1$ and vitamin A$_2$ and their aldehydes, retinal$_1$ and retinal$_2$, can exist in a number of different chemical and geometrical configurations corresponding to the *trans* to *cis* isomerization around the double bonds of these molecules (Figures 8.2, 8.3). For example, there are five possible geometric isomers of retinal, corresponding to rotation about the 9—10 carbon bond, the 11—12 bond, the 13—14 bond, and between bonds 9—10, 13—14 and between bonds 11—12, 13—14 (Table 8.1).

Of these geometric isomers it is the 11-*cis* retinal that complexes with the protein opsin in the visual pigment rhodopsin. The 11-*cis* retinal in rhodopsin was considered improbable because the stearic interference between the methyl group at carbon 13 and the hydrogen at position 10 would prevent the molecule from becoming entirely planar. But, in fact, Wald (1952, 1959) and his associates found that the functional isomer of retinal was the 11-*cis* retinal. The 11-*cis*-hindered configuration is the least stable of the possible isomers; it is the most easily formed upon irradiation and the most sensitive to light and temperature. The instability of

Retinals

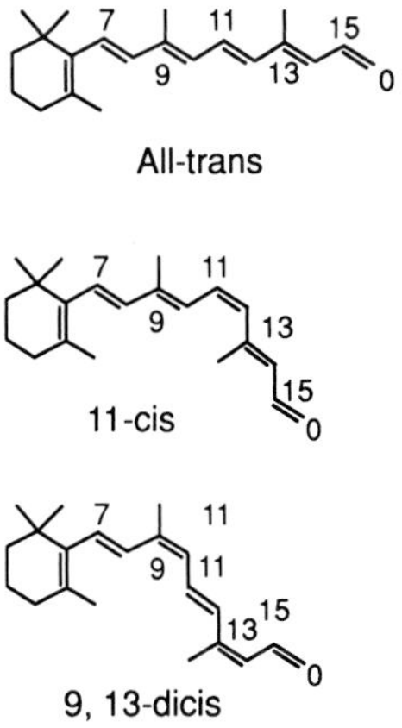

FIGURE 8.3   Isomers of retinals.

11-*cis* retinal, according to Wald, would explain its presence in rhodopsin, a molecule that is very unstable in the light but very stable in the dark.

Other geometric isomers, for example the 9-*cis* retinal, complex with opsin to form a series of photosensitive isorhodopsins, which have been found in small concentrations in the liver and in the blood but not in vertebrate eyes, and the 13-*cis* retinal is found in bacteriorhodopsin of halophilic bacteria.

RHODOPSINS

The present knowledge of the chemical structure, absorption spectra, and photo-chemistry of rhodopsins extracted from animal eyes is due to the pioneer studies of George Wald (1953, 1956, 1959), Dartnall (1957, 1962), and Crecitelli (1972, 1977).

Rhodopsins are retinal-protein complexes in which retinal is covalently bound to the protein opsin. The aldehyde (—CHO—) of retinal is attached to opsin via an ϵ-amino acid lysine in opsin by way of a protonated Schiff-base. The spectral sensitivity peak, spectral absorption maximum, is determined by the amino acid structure of the protein opsin and whether it has retinal$_1$ or retinal$_2$ in the rhodopsin complex. Opsin is *species specific* and is under genetic control (in the coding of the amino acid sequences in the synthesis of opsin).

> *The only way for a gene to code for an amino acid sequence of a protein is by means of its base sequences.*
>
> —FRANCIS CRICK, 1988, *What Mad Pursuit*

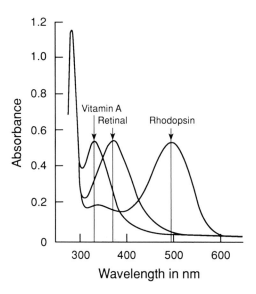

FIGURE 8.4   Absorption spectra of vitamin A, retinal, and rhodopsin.

There are now sufficient experimental data to show that the visual pigment rhodop-sins for photoreception and visual excitation are similar at the molecular level throughout animals.

Rhodopsins extracted from retinal photoreceptors are identified principally by their spectral absorption peaks as shown in Figure 8.4 for vitamin A, retinal, and rhodopsin. For the retinal rods, they are either ($retinal_1$ + rod opsin) around 500 nm or a porphyropsin ($retinal_2$ [dehydroretinal] + rod opsin) around 525 nm. Rhodopsins from the retinal cones are retinal + cone opsins whose rhodopsin absorption spectral peaks are around 455, 530, and 620 nm (Figures 8.5 and 8.6). Rhodopsins based on $retinal_1$ are from vitamin A_1, and porphyropsins based on 3-dehydroretinal are from vitamin A_2. In the light bleaching of rhodopsin, there is a transformation of the 11-*cis* retinal to the *all*-trans retinal (Figure 8.7).

Rhodopsins are found throughout land vertebrates whereas porphyropsins are found mainly in some teleosts, amphibians, and aquatic reptiles. Land vertebrates and marine fish characteristically possess the rhodopsin ($retinal_1$) system, while the porphyropsin ($retinal_2$) system is found in amphibians and fresh-water fish. Fish that migrate between freshwater and marine environments possess both $retinal_1$ and $retinal_2$ in their rhodopsins.

Invertebrate visual pigments

There are now sufficient experimental studies to show that the visual pigments of invertebrate eyes are also rhodopsins that are chemically similar to the vertebrate

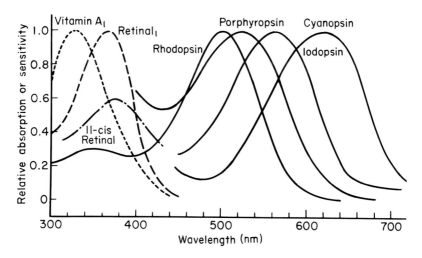

FIGURE 8.5   Absorption spectra of visual pigments. (From Wald, 1959; Dartnall, 1957.)

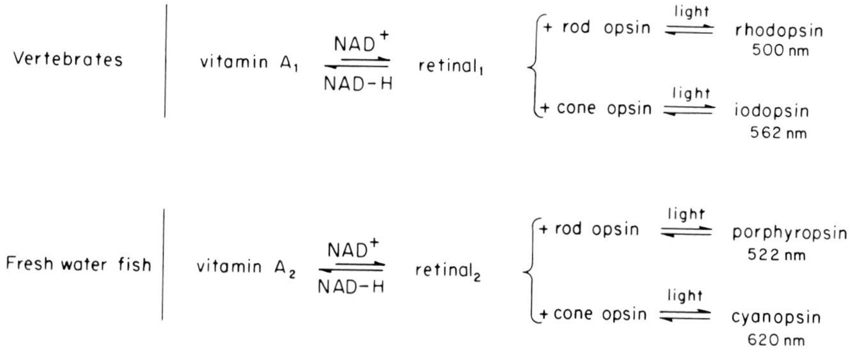

Vertebrates | vitamin A$_1$ $\xrightleftharpoons[NAD-H]{NAD^+}$ retinal$_1$ $\Big\{$ + rod opsin $\xrightleftharpoons{light}$ rhodopsin 500 nm / + cone opsin $\xrightleftharpoons{light}$ iodopsin 562 nm

Fresh water fish | vitamin A$_2$ $\xrightleftharpoons[NAD-H]{NAD^+}$ retinal$_2$ $\Big\{$ + rod opsin $\xrightleftharpoons{light}$ porphyropsin 522 nm / + cone opsin $\xrightleftharpoons{light}$ cyanopsin 620 nm

FIGURE 8.6   Visual pigments in the rods and cones, derived either from vitamin A$_1$ or vitamin A$_2$, indicating absorption maxima.

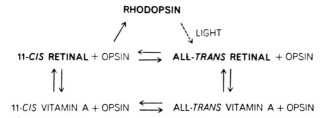

FIGURE 8.7   The light "bleaching" of rhodopsin and its regeneration back to rhodopsin.

107

rhodopsins. The invertebrate and vertebrate rhodopsins share a common chromophore, the 11-*cis* retinal, which is complexed to its protein opsin via the ε-amino group of lysine by way of a protonated Schiff-base. The photochemical process of invertebrate rhodopsin involves only the transformation of rhodopsin to metarhodopsin:

light

rhodopsin (11-*cis* retinal) ↔ metarhodopsin (*all-trans* retinal)

Retinal in metarhodopsin is isomerized back to 11-*cis* retinal in rhodopsin. The regeneration of rhodopsin from metarhodopsin takes place in light. This is a major characteristic of invertebrate rhodopsins and differentiates it from the vertebrate rhodopsins, which accomplish this regeneration in the dark (Goldsmith, 1975; Hamdorf, 1979).

This difference between invertebrate and vertebrate rhodopsins is due to the way retinal interacts with its specific protein opsin. Therefore, a continuous equilibrium between rhodopsin and metarhodopsin is established in the light.

Investigations of the absorption spectra and photochemistry of rhodopsins extracted from invertebrate eyes of insects, crustacea, and molluscs are reviewed by Goldsmith (1986), Hillman et al. (1983), and Tsuda (1987).

The universality of retinal

How universal is retinal and the rhodopsin system for photoreception in living organisms? The visual pigment rhodopsin and its chromophore retinal are found in the retinas of all animal eyes. Retinal was not found in eyeless animals until recently, and it was believed that only animals with eyes could synthesize retinal from its precursor molecules, vitamin A and β-carotene. However, we have already indicated that a rhodopsin photoreceptor system is found in halophilic bacteria, algae, and protozoa. Rhodopsin in these unicellular organisms is not for visual excitation, as in the retinal photoreceptors of animal eyes, but is a photoreceptor molecule for light detection and phototactic movement. For example, in the bacterium *Halobacterium halobium* and in the alga *Chlamydomonas,* a rhodopsin is the photoreceptor molecule associated with photoprocesses for phototactic movement and photosynthesis other than visual excitation. In these organisms, retinal is complexed to a protein through a Schiff-base with the amino acid lysine in the protein, as in rhodopsin. It was also indicated that retinal is more widely distributed in organisms—for example in photosensory cells in the skin, in pigmented neural cells, and in the brain pineal organ. Therefore, retinal has a more generalized function than was previously thought.

The fact that retinal is found in bacteria would indicate that its synthesis must have occurred very early in the history of life. At that time, there was much more ultraviolet radiation present in the environment. Evolving organisms, to protect their DNA from radiation damage—hence death—catalyzed the synthesis of reti-

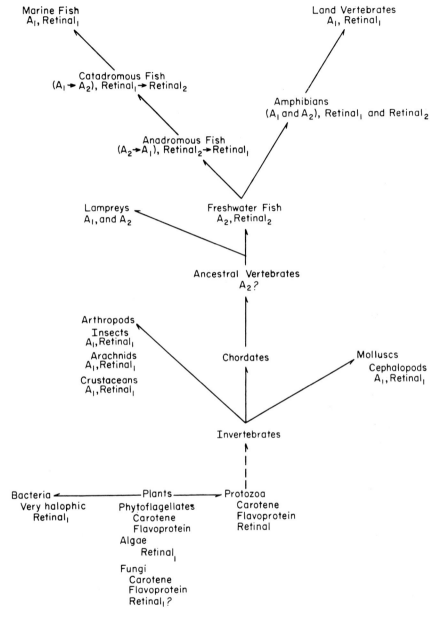

FIGURE 8.8 Phylogenetic relationship of vitamin A and retinal in photoreceptors. (Adapted from Wald, 1970, and added to by Wolken, 1975, 1986).

nal from its precursor molecule, β-carotene. Therefore, one can speculate that retinal, in its origin, was a protective mechanism against ultraviolet radiation damage.

As living organisms evolved and eyes developed, retinal complexed with a protein to form the visual pigment rhodopsin. Once incorporated in the retina of the eye, it was used for all visual systems that evolved, demonstrating natural selection at the molecular level (Wald, 1959).

The universality of retinal in the evolutionary development of photoreceptor systems in living organisms is indicated in Figure 8.8.

CONCLUDING REMARKS

The visual pigments of all vertebrate and invertebrate eyes are the rhodopsins. The rhodopsins extracted from vertebrate and invertebrate eyes indicate a range of absorption from the near ultraviolet through the visible into the infrared, 340 nm to beyond 700 nm.

All rhodopsins share a common chromophore, the 11-*cis* retinal, that is complexed with their specific protein opsin via the ε-amino acid of lysine in opsin by way of a protonated Schiff-base. The action of light on vertebrate rhodopsins is the release of 11-*cis* retinal from rhodopsin to the *all-trans* retinal and opsin. In all cases, the activated rhodopsin communicates its information with the rest of the cell/organism via a G-protein coupled sensory transductor cascade.

Although vertebrate and invertebrate rhodopsins are similar in chemical structure, there are differences in their photochemical intermediates. For invertebrate rhodopsins, unlike vertebrate rhodopsins, the photochemical process involves only the transformation of rhodopsin to metarhodopsin, and in the regeneration of rhodopsin, the *all-trans* retinal is isomerized back to 11-*cis* retinal, a process which takes place in the light and is reversible and continuous. The protein opsin in rhodopsin is species specific and is under genetic control, which determines the spectral sensitivity peak of the rhodopsin. Despite the independent evolutionary development of eyes and their visual pigments, rhodopsins are chemically remarkably similar.

Rhodopsins of vertebrate and invertebrate eyes and how they function for visual excitation are discussed in Chapters 9 through 12.

CHAPTER NINE

Vertebrate Eyes: Structure and Visual Excitation

There is much more to vision than meets the eye.

"The study of light originates from inquiry and speculation about the nature of vision."

— ABRAHAM PAIS, 1991, *Niels Bohrs' Times*

How we see is one of the oldest scientific questions asked by humans. Therefore, I have turned to the most highly evolved and developed eye, the vertebrate eye. The reason is that, of all of our special senses, vision is the most important to us, for about 40% of all sensory information about our world comes to us through the eye. The vertebrate eye is well-studied, and there is considerable knowledge of its structure, optics, and visual pigments. Therefore, it is of interest how our eye brings to us visual images of our world.

"The study of light originates from inquiry and speculation about the nature of vision."

— ABRAHAM PAIS, 1991, *Niels Bohrs' Times*

The anatomy of the mammalian (human) eye is shown in cross-section (Figure 9.1a). The eyeball is approximately spherical and houses the complete optical and photosensory apparatus. The optical system consists of the *cornea* (refractive index 1.336) and the crystalline *lens* (refractive index 1.437). The primary function of the cornea is to bend the incoming light to form the image on the retina. The lens is used to adjust the focusing of the cornea for near and far vision. The lens also acts as a filter by sharply cutting off the far edge of the ultraviolet region

a   **b**

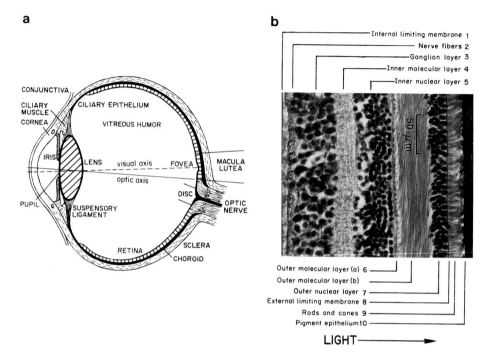

FIGURE 9.1   (a) Structure of the vertebrate mammalian eye. (b) Various cell layers in the human retina.

at about 360 nm. A variable aperture is provided by a contractile membranous partition, the *iris,* which regulates the size of the aperture's opening. The *pupil* is a hole formed by the iris through which light passes to the lens. Varying this opening directly affects the depth of field of focus much like a variable aperture camera. The retina's photoreceptor cells are the rod and cone cells, which, upon absorption of light, transduce the energy to chemical energy and to electrical signals that are transmitted via the optic nerve to the visual cortex in the brain.

The vertebrate eye is a refracting-type eye in which an image is formed by the refraction of light on one spherical surface that separates media of different refractive indices. The image formed is inverted, and its size is inversely proportional to the distance of the object. The refracting eye has the great advantage that image formation occurs through an integrative action, so that all rays falling on the eye from a given source are brought to a point of focus on the retina. Johannes Kepler around 1611 looked upon the eye as a *camera obscura,* or a refracting imaging instrument with the retina as a screen, and described the optics of this type of image formation, and René Descartes (1637) described how the eye accomplished imaging. In the beginning of the eighteenth century, William Molyneaux (1709) of Dublin published the first treatise on optics, showing diagrams of the projection of a real image in the human eye.

The combined optical and photosensory apparatus is paired in two symmetrically constructed and oriented eyeballs. As a result, a large section of visual space can be imaged on both retinas binocularly; that is stereoscopic vision.

This general description of the human eye and how it is structured for vision applies to all vertebrate eyes. All vertebrate eyes are modifications of this common plan.

The evolutionary phylogeny in the development of the vertebrate eye is discussed in greater detail in classic texts by Walls (1942), Polyak (1957), Sir Duke-Elder (1958), and Willmer (1960).

How the eye is structured for vision has been likened to a camera, but the processes that give rise to images of our world are far more complex. To better understand this complexity of the visual process, a review of the retinal photoreceptors, their molecular structure, photochemistry, and the photophysics that lead to signal transmission of images is in order.

THE RETINA

The development of the vertebrate eye involves principally the conversion of the cells in the wall of the optic cup into the retina. Some of these multiplying cells differentiate into the light-sensitive retinal rods and cones and others into the nerve cells. Although the original connection of the optic nerve with the embryonic brain persists throughout this process (as the optic stalk), the nervous connection of the retina with the brain is formed by the outgrowth of nerve fibers from the nerve cells of the retina through the optic stalk into the brain.

The vertebrate mammalian retina is a complex structure of ten cell layers that are closely attached to the pigment epithelium (Figure 9.1b). The first four cell layers of the retina constitute the neuroepithelial cells and are the neurons of the first order. The remaining layers are considered the cerebral portion, where there exists a complex arrangement of nervous elements resembling those of the central nervous system in structure and function; in essence the retina represents an outlying portion of the central nervous system. The fifth cell layer (internal nuclear layer containing the bipolar, horizontal, and amacrine cells) and the sixth layer comprise the neurons of the second order. The seventh and eighth cell layers make up the neurons of the third order, which pass centripetally to the primary optic center (the lateral geniculate) of the metathalamus. The rod and cone cells comprise the ninth cell layer and are closely attached to the pigment epithelium, the tenth cell layer. The nervous cell layers of the retina are the rod and cone cells, the bipolar cells, and the ganglion cells. Light passes through all the cell layers of the retina before it reaches the photoreceptor rod and cone cells. Neural information then flows back through the layers of the retina from the photoreceptors, through the bipolar cells, to the ganglion cells. The horizontal and amacrine cells modify the visual signal as it passes through the bipolar cell layer. From the ganglion cells, visual information is passed on to the primary optic center (lateral geniculate) of

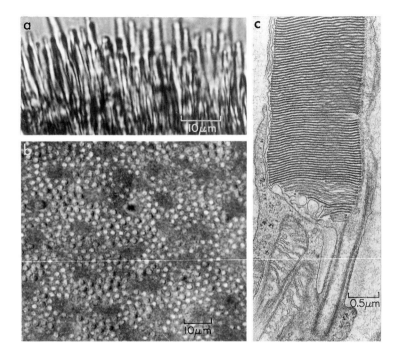

FIGURE 9.2   The human retinal photoreceptors. (a) Retinal rods and cones extending from a bend in the retina. (b) Surface view of retinal rods and cones. Note array of rods (rhabdomlike arrangement). (From Wolken, 1966, p. 23.) (c) Electron micrograph of a human retinal rod showing the outer segment lamellae and inner segment of the retinal cell. (Courtesy of Dr. T. Kuwabara, National Institutes of Health, Bethesda, Maryland.)

the metathalamus. This inversion of the retina in vertebrates is the result of the development of the eye as an outgrowth of the embryonic brain. Next to the retina is the choroid coat, a sheet of black melanin-pigmented cells that absorbs extra light and prevents internally reflected light from blurring the image.

The retinal photoreceptors, the rods and cones, are arranged in a single-layered mosaic in a rhabdom-like array (Figure 9.2a,b), connected with a highly developed system of interconnecting neurons. In the human retina, there are about 1×10^8 retinal rods and about 7×10^6 retinal cones. Toward the center of the human retina there is a depression, the fovea, which is the fixation point of the eye and where vision is most acute. It contains mostly cones, of which there are 4×10^3. The rods become more numerous as the distance from the fovea increases. The fovea and the region just around it, the macula lutea, are colored yellow; they contain a plant carotenoid pigment, xanthophyll. The structural architecture of the retina, the electrophysiology of the neural cell layers, and their relationship to vision is reviewed by Dowling (1987) and Masland (1986).

RODS AND CONES: PHOTORECEPTORS FOR PROCESSING INFORMATION

The search for a molecular basis for visual excitation has led to the structural study of retinal photoreceptors. The retinal photoreceptors, the rod and cone cells, are specialized for photoreception and visual excitation. In each retinal cell, all the visual pigment, rhodopsin, is contained in a rod- or cone-shaped outer segment (OS). The index of refraction of the rod OS is 1.41 and that of the cone OS is 1.39. The inner segment (IS) is the retinal cell body which possesses a nucleus, mitochondria, and other cellular organelles typical of animal cells. Clusters of rod cells are connected to optic nerve fibers, whereas the cone cell is connected to a single optic nerve fiber. The rods and cones are connected with a highly developed system of connecting and interconnecting neurons.

Developments in electron microscopy and x-ray diffraction have made possible the visualization of the microstructure of the retinal rods and cones in molecular dimensions. Electron microscopic studies have clearly established that the vertebrate retinal rod OS are double-membraned, lipid-protein discs of the order of 200 Å in thickness, that each membrane of the disc bilayer is from 50 Å to 75 Å in thickness, and that these membranes are interspaced by water, enzymes, and dissolved salts.

In the rod OS, the discs appear as flattened plates that are piled up with no connection to the cell plasma membrane (Figure 9.2c). In the cone OS, the membranes are continuous with the plasma cell membrane. The retinal rod and cone structures are schematically shown in Figure 9.3. In the retinal rod, interconnection from the OS through the IS occurs through a cilium (or flagellum). The structure of this cilium distinctly shows the characteristic nine fibrils found in cilia and flagella of plant and animal cells.

The connecting cilium may be a crucial factor in the cells' functional chemistry. For at one end the OS is a highly ordered, photosensitive matrix containing all the rhodopsin and at the opposite end is the cell body with a mass of mitochondria whose enzymatic action provides the oxidation-reduction chemical reactions and hence energy transfer.

Much of what we have learned about the structure, chemistry, and photochemistry of rhodopsin of the retinal rods has come primarily from studies of amphibian and bovine retinal rods. The amphibian retinal rod OS can be severed from the retina simply by shaking in frog Ringer solution and are easily observed with the light microscope. In examining the frog *Rana pipiens,* the retinal rod OS are unusually large, about 6 μm in diameter and about 60 μm long (Figure 9.4). They appear to be highly refractive, indicating an ordered microstructure. Electron microscopy of fixed and sectioned retinal rods shows that all vertebrate rods consist of discs, or double-membraned lamellae. A cross-sectional view of freshly fixed and sectioned frog rods shows a cylinder with scalloped edges and fissures extending into the rod, so that it is divided into fifteen to twenty irregular wedges

Retinal Rod

Retinal Cone

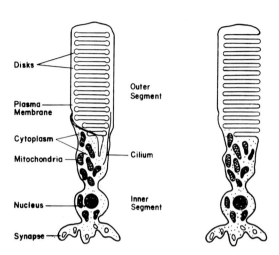

Disks

Outer
Segment

Plasma
Membrane

Cytoplasm

Mitochondria

Cilium

Nucleus

Inner
Segment

Synapse

FIGURE 9.3   Schematic of retinal rod and cone cells, indicating the structural difference in that rod OS are formed of discs, while cone OS are pleated structures of the plasma membrane.

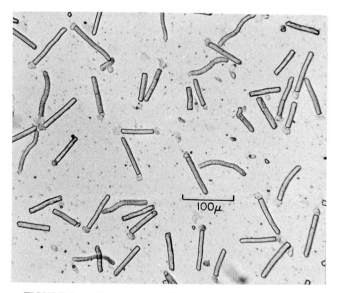

100μ

FIGURE 9.4   Frog retinal rods isolated from the retina.

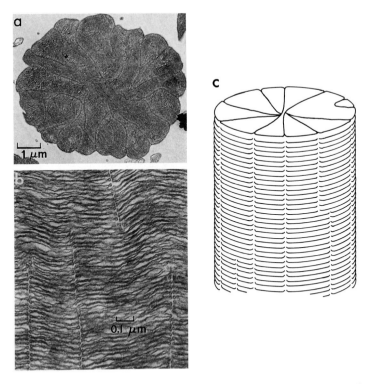

FIGURE 9.5 Frog retinal rod. (a) Cross-section of outer segment. (b) Longitudinal section. (c) Schematic of rod outer segment structure. (After Wolken, 1971.)

(Figure 9.5). Longitudinal sections reveal that these wedges are structured of rodlets of about 1 μm in diameter within the rod structure.

Detwiler (1943) observed that the retinal rods expanded during illumination and contracted after dark adaptation, and he referred to this behavior as a photomechanical mechanism. Therefore, it was interesting to reinvestigate this behavior in relation to the structural changes in the retinal rods. The frog retinal rod OS were isolated in red light, immersed in the vitreous fluid of the eye (to prevent osmotic shock), and illuminated. Structural changes were photographed through the microscope (Figure 9.6, 1–4) while simultaneous spectral absorption changes were recorded with the microspectrophotometer (Figure 9.7, 1–4) (Wolken, 1966, 1975). It was found that the rod expanded to almost twice its length without changes in its diameter; this expansion was accompanied by the "bleaching" of rhodopsin, as shown in Figures 9.6 and 9.7 (Wolken 1966, 1975). When these rod OS were restored to darkness and 11-*cis* retinal was added, some retraction of the rod and regeneration of rhodopsin occurred. The addition of ATP resulted in further rod retraction, and when illuminated, the rod expanded again. The rod OS

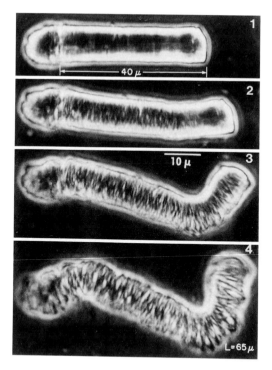

FIGURE 9.6 Freshly isolated frog retinal rod (1), irradiated with white light (2–4). Note change in length and structural changes.

behaves much like a spring, a kind of "Jack-in-the-box" effect which is triggered by light.

These structural changes in the rod OS were then examined by fixing the freshly isolated rods before and after illumination and scanning them with the electron microscope. The electron micrographs of the illuminated rods confirmed that OS disc membranes within the rod expanded. These observations suggest that light energy is transduced to chemical and mechanical movement and that the process is reversible. The behavior of the retinal rod OS is much like the expansion and contraction of muscle fibers upon stimulation. Support for this hypothesis comes from the fact that retinal cells, as with all cell skeletal structures, are formed of muscle-like protein filaments. Experiments have identified one of the structural cytoskeletal proteins of the retinal rod—the muscle protein F-actin (Del Priore et al., 1987).

PHOTOCHEMISTRY OF RHODOPSIN

The mechanism of visual excitation that takes place via the retinal photoreceptors must be examined with reference to photophysics and photochemistry. That is how light energy is transduced to chemical energy and to electrical signals from which

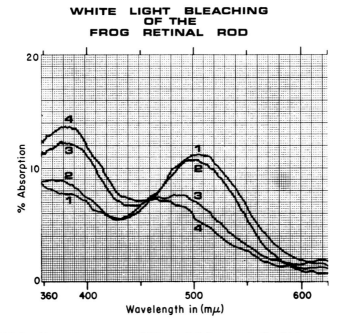

FIGURE 9.7   Absorption spectra of Figure 9.6 (curves 1–4) which accompany these structural changes (i.e., rhodopsin curve 1 to retinal curve 4). (Obtained with a micro-spectrophotometer.)

the image is recreated in the visual cortex of the brain. The photochemistry of rhodopsin has been actively studied, and there is some understanding of how it is related to this process.

Upon light absorption, there is a conformational change in the shape of rhodopsin which is accompanied by the change from the 11-*cis* retinal to the all-*trans* retinal in the membrane, and photoexcitation occurs. In the dark, the reverse process occurs and rhodopsin is restored to its original state in the membrane. When light is absorbed by rhodopsin, it "bleaches," changing in color from reddish-purple to yellow, and the absorption spectrum is shifted from around 500 nm to around 370 nm. The bleaching of rhodopsin by light can be followed spectroscopically, as shown in Figure 9.8 where curve 1 is the absorption spectrum of frog rhodopsin that absorbs around 500 nm and where curves 2–7 show the displacement of the major rhodopsin peak around from 500 nm to around 370 nm, that of all-*trans* retinal. In the ultraviolet region of the spectrum, the absorption peak, around 280 nm, is that of the protein opsin. There is no significant change in the opsin absorption peak during the light bleaching process. In the intact eye, the lens limits short wavelength light (<400 nm) from reaching the retina. In the light bleaching of rhodopsin, the 11-*cis* retinal is isomerized to the all-*trans* retinal. To resynthesize rhodopsin the all-*trans* retinal is isomerized back to the 11-*cis* retinal to recombine with opsin to form rhodopsin (Figure 8.7).

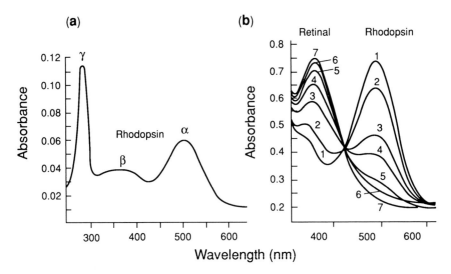

FIGURE 9.8   (a) Spectrum of frog rhodopsin (extracted in 4% tergitol); the absorption peak around 280 nm is due to the protein opsin. (b) Spectral changes upon light bleaching (curves 2–7). (Compare to Figure 9.7.) (From Wolken, 1975.)

This photoprocess, by which the 11-*cis* retinal is uncoupled from opsin to the all-*trans* retinal, proceeds in a series of intermediate photoproducts. The spectral identities of these intermediates are still being investigated. In this photoprocess, the only light-catalyzed step is a sequence of events in which the protein bound retinal forms a high-energy photoproduct that undergoes thermal decay through a series of intermediates. The intermediates that have been identified have distinct absorption spectral peaks. Upon light absorption the first photoproduct, bathorhodopsin (formally prelumirhodopsin), has λ max about 543 nm and is stable below −150°C. On warming in darkness, the bathorhodopsin bleaches over a sequence of intermediates, each of which represents a stage in the step-wise opening up of the tertiary structures of opsin to lumirhodopsin, stable between −140°C and −40°C; to metarhodopsin I, stable from −35°C to −15°C; to metarhodopsin II, reasonably stable between −10°C and 0°C; and finally with further warming to the more stable all-*trans* retinal from opsin (Figure 9.9). To complete the cycle, the all-*trans* retinal isomerizes back to 11-*cis* retinal isomer, which recombines with opsin to again form rhodopsin. The reaction is spontaneous, and therefore opsin may be looked upon as a retinal-trapping enzyme, removing free retinal from the mixture and causing the production of additional retinal from vitamin A to maintain the necessary equilibrium.

The photochemical intermediates of rhodopsin and photophysics that take place in retinal photoreceptors of the eye are being investigated continuously, and these recent studies are discussed in the reviews by Birge (1990), Becker (1988), Lanyi (1992), Lewis and Del Priore (1987), and Pugh and Cobbs (1986).

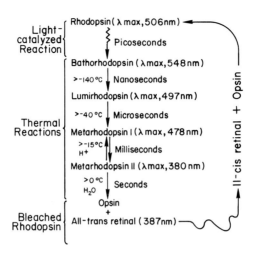

FIGURE 9.9   Rhodopsin (bovine); photochemical intermediates in the process of light and thermal bleaching reactions of rhodopsin.

G-proteins and visual excitation

The absorption of light by rhodopsin in rods and cones involves more than their photochemistry to bring about visual excitation. The mechanism includes an elaborate series of photoprocesses by the rod and cone cells where the light energy absorbed is transduced to chemical and to electrical signals that are carried via the optic nerve to the visual cortex in the brain. Visual excitation requires a transmitter molecule that changes the electrical potential of the cell membrane at the receptor sites. The discovery that the enzyme cyclic *guanosine monophosphate* (GMP) is a second messenger that changes the electrical potential of the receptor site in the membrane indicates that it also participates in the process of visual excitation. GMP does not absorb light but is activated by the interaction with bleached rhodopsin. This results in a light-triggered cascade, a very rapid hydrolysis of cyclic GMP, by phosphodiesterase. During this process, light reduces the inflow and the membrane current. Thus, cyclic GMP is an internal second messenger that links the photobleaching of rhodopsin to the electrical response of the retinal cell (Ross, 1988). Many features of the photoreceptor signal transmission pathways are not completely understood and are being actively investigated.

The retinal rod molecular structure

The vertebrate retinal rod OS structure consists of double membrane discs about 200 Å in thickness, with each membrane of the disc being about 50 Å in thickness. The rhodopsin molecules are intimately associated with the disc membranes.

Knowing the concentration of rhodopsin molecules and measurements of width, length, and number of disc membranes per retinal rod OS, the cross-

TABLE 9.1 Comparative Composition of Proteins and
Lipids in Retinal Rod OS[a]

	Dry weight (%)	
	Cattle	Frog
Total lipid	38.15	40.6
Total protein	61.85	59.4
	Total lipids (%)	
Phosphatidylethanolamine	38.5	25.2
Phosphatidylserine	9.2	9.5
Phosphatidylcholine	44.5	49.4
Sphingomyelin	1.3	1.8
Other phospholipids	6.5	9.2

[a] Data taken from various sources and averaged.

sectional area of the rhodopsin molecule can be calculated (Wolken, 1975). To
calculate cross-sectional area and the diameter of the rhodopsin molecule, several
assumptions are necessary. These are: rhodopsin is associated with the lipid bilayers of the disc membranes, and rhodopsin molecules oriented in the membranes
are nearly parallel to the surface of the membrane as a monomolecular layer. These
assumptions are supported by structural and chemical analyses showing that rhodopsin accounts for 60% and total lipids for about 40% of the weight of the rod OS
(Table 9.1).

The cross-sectional area A that would be associated with each rhodopsin molecule can be expressed by

$$A = \pi D^2/4P$$

where D is the diameter of the retinal rod and P is the number of rhodopsin
molecules in a single monolayer. In the equation for the maximum cross-sectional
area for each rhodopsin molecule, P is replaced by $N/2n$, where N is the rhodopsin
concentration in molecules per retinal rod and n is the number of disc membranes
per rod.

$$A = \pi D^2 n/2N$$

The concentration of rhodopsin in the frog retinal rod is 3.8×10^9 and for the
bovine retinal rod is 4.2×10^6 molecules (Tables 9.2–9.4). Inserting the data in
this equation, the cross-sectional area calculated for bovine and frog rhodopsin are
2500 and 2620 Å², respectively (Table 9.4), which means the diameter of the
rhodopsin molecule would be about 50 Å (Wolken, 1975). If the rhodopsin molecule is symmetrical, the diameter would be around 40 Å (Wald, 1954). The most
probable distance between rhodopsin molecules in the monolayer (Chabre, 1975)
is calculated to be about 55 Å, if they are aligned with their long axis perpendicu-

TABLE 9.2   Vertebrate Retinal Rod OS

Animal	Diameter (μm)	Length (μm)	Volume (cm^3)
Frog	5.0	55	1.1×10^{-6}
Perch	1.5	40	6.2×10^{-11}
Chicken	3.5	35	3.4×10^{-11}
Cattle	1.0	10	7.5×10^{-12}
Monkey	1.3	22	2.3×10^{-11}
Human	1.0	30	1.6×10^{-10}

From Wolken (1971). Average measurement.

TABLE 9.3   Retinal Rod Volume and Rhodopsin

Vertebrates	Retinal rod	
	Average volume (cm^3)	Concentration of rhodopsin molecules
Frog (*Rana pipiens*)	1.1×10^{-6}	3.0×10^9
Çattle	7.5×10^{-12}	1.0×10^6
Human	1.6×10^{-10}	1.0×10^7

From Wolken (1971).

TABLE 9.4   Retinal Rod OS Structural Data

Animal	Average diameter, D (μm)	Thickness of disc, T (Å)	Number of lamellae per rod, n	Rhodopsin molecules per rod, N	Calculated cross-sectional area of rhodopsin (Å^2)	Calculated diameter of rhodopsin molecule, d (Å)	Calculated molecular weight, M
Frog	5.0	150	1000	3.8×10^9	2620	51	60,000[a]
Cattle	1.0	200	800	4.2×10^6	2500	50	40,000[b,c]

[a] From Wolken (1975).

[b] Calculation based on a lipoprotein, density 1.1, gives a molecular weight of 32,000 daltons.

[c] Abrahamson and Fager (1973) indicate molecular weight of 35,000–37,000 daltons.

123

lar to the disc membrane. These calculations for the cross-section of rhodopsin indicate that there would be sufficient space to accommodate all the rhodopsin molecules on the disc membranes. Furthermore, the synthesis of rhodopsin molecules is directly related to the number of disc membranes in the retinal rod OS, indicating the relationship of growth to development at the molecular level for the photoreceptor membranes.

An estimate of the rhodopsin molecular weight M can also be calculated from the data. Where D is the diameter, T the thickness of the disc membranes, s the density (taken as 1.3 for a protein), 1 Avogadro's number, n the total number of membranes, and N the number of rhodopsin molecules, the molecular weight is then obtained from:

$$M = \pi D^2 Tsln/4N$$

The molecular weight calculated from this equation for frog rhodopsin was found to be 60,000 and for bovine rhodopsin 40,000 daltons (Table 9.4). This compares well with the bovine rhodopsin molecular weight of 40,000 daltons calculated by Hubbard (1954), 37,000 by Abrahamson and Fager (1973), and 38,000 by Chabre (1975). If, in calculating M, the density of a lipoprotein (1.1) is used for s, the molecular weight would be reduced by 20%, resulting in a molecular weight of 32,000 daltons for bovine rhodopsin and 48,000 daltons for frog rhodopsin (Wolken, 1966, 1975). These molecular weights correlate well with the predicted size inferred from the protein opsin amino acid sequences.

The molecular structure of a retinal rod OS is schematized in Figure 9.10. A small area is enlarged to show how the rhodopsin molecules are associated in the lipid bilayer of the rod disc membranes in which the polypeptide chains enter into the aqueous environment of the membrane as seven α-helices. Retinal is bound to a lysine amino acid residue in the carboxyl retinal helix, is held in a pocket, as depicted in Figure 9.10, according to Unwin and Henderson (1975), and may hold for other photoreceptor membranes as well. This molecular model may be very close to reality, for x-ray diffraction studies of the frog rod OS seem to support such a model (Blasie, 1972). Rhodopsin has not yet been crystallized, so we do not presently have a three-dimensional molecular model of the rhodopsin molecule and how it is molecularly associated with the retinal rod membranes.

CONCLUDING REMARKS

The vertebrate visual photoreceptors (rods and cones) are structured of lipid-protein membranes into lamellae (discs for rods), which, in cones, originate in the retinal cell plasma membrane. Therefore, the photoreceptor membranes are chemically and structurally similar to cellular membranes to which the visual protein become molecularly associated.

Rods and cones are sensitive to light, temperature, pressure, and electrical stimuli. These are the very properties of a liquid crystal, and, therefore, we have considered them as liquid crystalline structures (Brown and Wolken, 1979).

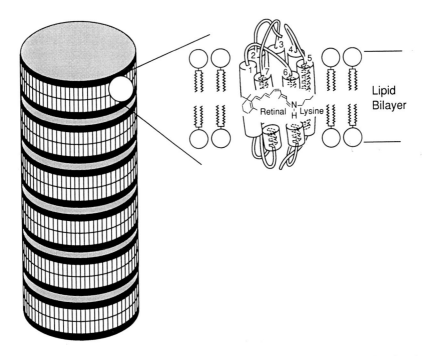

FIGURE 9.10 Schematic model for molecular structure of retinal rod (OS) showing the possible molecular geometry of retinal with opsin (rhodopsin) in the membrane of the retinal rod.

Another consideration is that rods and cones are optical devices for transmitting light through the retina. If so, how do they function? Enoch and Tobey (1981) investigated the retinal rod of an optical device and found that the rod was structured much like a fiber optic light guide. This analogy is due to the fact that the retinal rod has a high index of refraction (1.41) and is surrounded by a lower index of refraction (1.34), similar to fiber optic filaments. In the retina, many rods are associated together—fiber optic bundles, light pipes, wave guides—to maximize light transmission through the retina.

The described vertebrate eyes are then highly specialized for vision and are adapted for serving the brain. However, vertebrate eyes are limited to the visible wavelengths of light, the retinal photoreceptors lack of screening pigments, so there is no color enhancement. Mammalian eyes are unable to detect polarized light in the environment (exceptions are birds and fish, Waterman, 1989). These limitations in the vertebrate visual system were overcome in the evolutionary development of arthropod (insects and crustacea) eyes, as will be described in discussions to follow.

CHAPTER TEN

Bird and Fish Eyes

We have previously discussed vertebrate eyes and how they are structured for vision. Here, I would like to indicate how birds and fish have eyes that are structured differently from those of land vertebrates.

In the evolutionary development of animal phyla (Figure 7.1), the chordates (fish, birds, and mammals) arose via separate pathways. Ancestral reptiles gave rise to modern reptiles and birds. Birds therefore have a reptilian origin, and the avian eye has many structural features in common with the reptilian eye.

THE EYES OF BIRDS

There are more than 10,000 species of birds, and a variety of photobehavior is found among them. Most birds are diurnal, although some are nocturnal and some are even amphibious.

The avian eye is considered one of the most highly developed visual organs in the animal kingdom. Birds have greatly improved visual acuity compared to terrestrial vertebrates, which is evident in their ability, in mid flight, to image creatures in the air, on land, or on the surface of water.

In observing the eyes of birds, one is struck by the size and shape of their eyes compared to the size of their heads (Figure 10.2). With large eyes, birds are able to obtain a greater field of view. Most birds have *flattened*-shaped eyes that give a broad field of vision; others have *globose*-shaped eyes, common to birds that hunt in daylight; and nocturnal species have *tubular*-shaped eyes. A schematic of the avian eye is shown in Figure 10.1; the structures identified are the cornea, lens, aqueous humor, vitreous, and retina, which are common to all vertebrate eyes. In addition there is a unique structure, the *pecten*. It arises from the linear optic disc and projects freely into the vitreous of the avian eye as a convoluted, accordion-

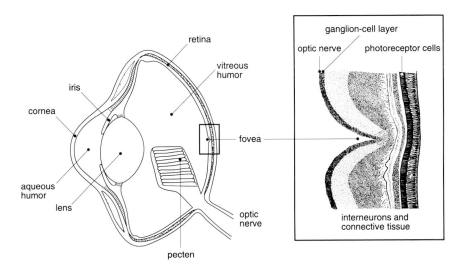

FIGURE 10.1 Cross-section of the avian eye identifying the various eye structures. Insert shows retinal cell layers in the area of the fovea.

pleated, lamellar structure varying in size and complexity. The size of the pecten and the number of lamellae vary and do not necessarily coincide with the size of the avian eye; these appear to be directly related to the degree of illumination the bird receives. Active, diurnal birds with high visual acuity are found to have a larger pecten with more lamellae. Nocturnal birds have a smaller pecten with fewer lamellae, and consequently they have poorer vision.

The function of the pecten structure in the avian eye is believed to improve the bird's visual acuity and to detect moving objects. In doing so, the pecten casts a shadow upon the retina that is not completely symmetrical, is significantly large, and this shadow influences the response of the retinal photoreceptors. The shadow created by the pecten structure is thought to increase the retina's sensitivity so that the eye is better able to detect the movement of objects in its surroundings (Menner, 1938). Other functions that have been suggested for the pecten structure in the

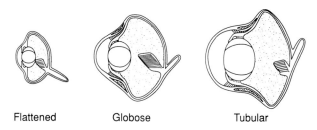

FIGURE 10.2 Various shapes of avian eyes.

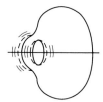

FIGURE 10.3   Cornea and lens changes in response to light intensity.

eye include improved light absorption, secretion, and even heat exchange (Meyer, 1986).

The avian eye is a refracting-type eye. The lens is biconvex, and the greatest refraction occurs at the surfaces of the cornea and the lens. The image on the retina, as in the mammalian eye, is inverted. The ability to focus the eye on objects at various distances, or *accommodation,* is well-developed in birds. It involves the simultaneous alteration of the power of each of the eyes by changing the curvature of the cornea and lens (Figure 10.3) and thus permits birds to have a greater focusing range than other vertebrates.

The retina

The avian retina is structured, as in all vertebrates, of rods and cones but with numerous single and double cones. Nocturnal birds have a greater number of retinal rods compared to the number of cones. The fovea, a depression in the retina, is a region of high visual acuity. Light is refracted at the surface of the fovea, which tends to magnify the image projected on the photoreceptor cells. Snyder and Miller (1978) described in falconiform eyes a telephoto lens system (Figure 10.4). They have shown that the presence of a spherical depression in the deep fovea acts like a negative lens component in a telephoto lens system. The focal length of the bird's dioptrics can then exceed the axial length of the eye, providing a relatively large image and high resolving power in a localized region of the retina. The disadvantage of the telephoto lens system is that it has a narrow field of view.

Associated with cones are pigmented oil globules that lie in the IS adjacent to the OS. They range in color from yellow, orange, red, to colorless. In the cones of nocturnal birds, the oil globules are faintly colored or colorless. They function as cut-off filters of light reaching the OS, making cones less sensitive to light transmitted in the ultraviolet to red wavelengths. The colorless globules are transparent and transmit all wavelengths of light. Other functions that these pigmented oil globules have in the retinal cones of birds regarding their spectral sensitivity and color vision are discussed in Chapter 11.

Nocturnal birds have a greater number of retinal rods compared to diurnal birds. In addition to rod rhodopsin, three different spectral absorbing visual pigments are

Corneal Lens

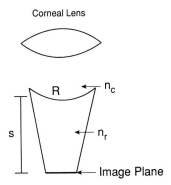

FIGURE 10.4   Telephoto lens system of falconiform eyes. (From Snyder and Miller, 1978). n_c and n_r are the refractive indices of the medium to the right and left, respectively, of the spherical surface. The radius, R, of curvature of the surface and the distance, s, from the apex of the spherical surface to the image plane.

found in cones that absorb in the blue, green, and red. Behavioral studies of hummingbirds and pigeons show sensitivity to wavelengths around 350 to 380 nm, indicating an ultraviolet light photoreceptor pigment as well in the cones. In aerial birds, the retinal cones may have as many as five, or, if the retinal rods are included, six, different absorbing visual pigments that can cover the whole visible spectral range from the near ultraviolet into the red (Ohtsuka, 1978, 1985). Thus, the cone visual pigments enable birds to have color vision.

The evolution of birds and the development of the avian eye, their structure, and their visual acuity are discussed in Walls (1942), Rochon-Duvigneaud (1943), and Duke-Elder (1958). More recent studies of the avian eye can be found in Waldvogel (1990), and Zeigler and Bischof (1993).

THE EYES OF FISHES

Fish are more diverse than birds or land vertebrates, with more than 25,000 different species estimated to exist. Many fish dwell close to the water surface, others are mid-water or deep-sea dwellers, and some are even amphibious. Depending on their habitats, they exhibit considerable variability in their visual behavior. Fish eyes are similarly structured to those of terrestrial vertebrates (Figure 9.1), but there are differences in their optics, types of retinal photoreceptors, and visual spectral sensitivities (Nicol, 1989).

Fish, in water, require the optics of their corneas and lenses to function differently from those of terrestrial animals. Under water, there is no air/cornea interface, and the eyes derive no benefit from the refractive power of their corneas, due to the fact that the index of refraction of water is 1.33. To compensate for the lack of the refractive power of the cornea, fish evolved a large spherical lens to capture as much light as possible.

> *Thus the lens of the fish eye was more spherical than the lens in the eye of land vertebrate because each was adapted to the refractive index of the medium, water or air.*
>
> —CHARLES DARWIN, quoted in R. W. CLARK, 1989,
> *Survival of Charles Darwin*

The lens

Fish lenses are spherical, rigid, non-elastic, and have a dense core that has a high refractive index. Such spherically shaped lenses are ideally designed to maximize light gathering, which is needed for imaging in an aquatic environment.

James Clerk Maxwell (1861) had a lifelong interest in optics, the eye, and vision. He is associated with the development of the tricolor theory of color vision. There is an interesting legend related to Maxwell—that while eating kippers, he contemplated the optics of the crystalline lens of the herring eye. In search of a perfect imaging lens, Maxwell's investigation of geometric optics led to a beautiful discovery, published in 1853, of the imaging properties of the "fish eye" lens. Maxwell postulated that the fish spherical lens had a refractive index gradient. He first published this in an obscure Irish journal, though it was later republished in his collected work *Some Solutions of Problems*. An interesting historical account of how Maxwell arrived at the optics of the "fish eye" lens is given in Pumphrey (1961).

Matthiessen (1886) examined the optics of fish eye lenses in a wide variety of different species of fish. He found the ratio of focal length to radius to lie within the limits of 2.5 and 2.6 and realized that so short a focal length was only possible if the refractive index of the lens fell radially from the center to the periphery. Matthiessen originally proposed that the refractive index n would vary with the distance r from the lens center as $n^2 = a - br^2$, where a and b are constants. The lowest possible refractive index in tissues of aquatic animals is that of water (1.33), and this value sets the lower limit for the refractive index at the periphery of the lens. It also explains why there must be a high refractive index at the center of the lens (1.51 to 1.53) radically decreasing continuously and symmetrically in all directions to the periphery of the lens (1.33). He postulated that such an index gradient of refraction would correct for spherical aberration in the lens. If the lens were homogeneous, the index of refraction would be 1.66, but according to Land (1981) and Fernald and Wright (1983), the index of refraction of the center of the lens is 1.51 to 1.53, just as Matthiessen predicted.

Much later, and in a context having nothing to do with the optics of fish eyes, Luneberg (1944) computed the general function and provided the theoretical basis for describing an ideal refractive index for a spherical lens. That is, a bounded spherical lens, aplanatic and free of spherical aberration, would have a refractive index gradient steep enough to form an image at the rear surface of a spherical lens. The focal length of such a lens would depend on the steepness of the refractive index gradient (Fletcher et al., 1954). Such an ideal spherical lens is known as the "Luneberg lens."

The spherical fish eye lens has a refractive index gradient increasing from its periphery, where it is the same as that of water (1.33), to its center, where it is greater than 1.5 (Fernald and Wright, 1983). Thus, there are no defined surfaces of refraction that can be used to calculate the bending of incident rays. Instead, the refractive index gradient forces central rays to slow down more than peripheral rays because the ray path in the lens is smoothly curved. Such a spherical lens with a graded index evolved in aquatic animals to function as a light concentrator in the optical axis. It is generally free of spherical aberration and functions equally well in all directions, providing fish with excellent imaging over a wide visual field.

The lenses of many fish that live in shallow water are pigmented yellow, and in some of these fish the cornea is yellow as well. Deeper-sea fish, *elasmobranchs* (skate, dogfish, and shark), do not have yellow-pigmented lenses (Zigman, 1991). However, in the deep sea at depths of 500 meters, *mesophlagic* fish, *Angropeleus affinis* for example, have a lens that is bright yellow, and its spherical absorption peaks at about 405, 430, and 460 nm (Sominya, 1976). The yellow lenses of these fish act as cut-off filters for the light of short wavelengths, like the yellow oil globules in the retinal cones of birds. In fish at greater depths, a yellow optical filter would serve to counter camouflage coloration of animals swimming above them (Lythgoe, 1979). Furthermore, the pigmented yellow lens eliminates chromatic aberration of the image and reduces scattered light as well as glare (Zigman, 1991). Whether the yellow lens acts only as a cut-off filter and to increase visual acuity of these fish is not presently known.

The retina

The fish retina is structurally similar to that of land vertebrates (Figure 9.1b). In some species of deep-sea fish there are double or multiple retinas, thus increasing the number of photoreceptors available to ensure light collection from their environment to excite the eye. The fish retinal photoreceptors are rods and various types of cones. There are single cones, double cones, and twin cones, which in some fish are combined into one cone. In many fish retinas, it is difficult to distinguish morphologically cones from rods. Not only are cone types variable but so are their arrangements within the retina. The different cone types may be randomly arranged or organized into rods or square mosaics, a feature of many teleost retinas (Layll, 1957; Engström, 1963).

ELASMOBRANCHS

According to Walls (1942) and Romer (1955), the lowest of the vertebrates are the cyclostomes, which include hagfishes and lampreys. Just above the cyclostomes are many types of true fish whose eyes are more specialized. The oldest of these fish are elasmobranchs, whose modern representatives are sharks, dogfish, skates, and rays. These are cartilaginous fish, which are believed to be descendants of Devonian forms that evolved from placoderm-like ancestors. Therefore, we can look for evidence that the elasmobranch eye developed from these more primitive ancestors and foreshadowed the development of the vertebrate eye.

The retina of elasmobranchs contains visual photoreceptors that are undifferentiated, i.e., no distinct rods or cones. However, observations of the lemon shark (Gruber et al., 1963) indicate that the retina has both rods and cones. The cones are characterized as being short, with tapering OS. They have pyramidal IS and are free of oil globules. Some elasmobranch retinas, for example the skate's, do not have cones (Dowling and Ripps, 1970). Most of the visual photoreceptor cells of the shark retina are rods, but there are other photoreceptors that have a greater diameter. Although these have been identified as cones, they may be modified rods. This visual cell morphologically resembles a rod, but due to the organization of its inner layers it is capable of functioning like a cone. Nonetheless, while sharks do distinguish changes in brightness, there is no evidence that they perceive color. Therefore, only one type of visual cell, resembling a rod more than a cone, is indicated in the elasmobranch retina.

In view of the fact that the sense of olfaction is so well-developed in sharks, one may postulate that their vision has evolved to be especially useful in locating nearby objects. Once prey is located from a distance by olfaction, the shark must depend on a fairly acute visual image. While the shark eye is not well-suited for seeing objects in sharp detail, it can distinguish moving objects from their backgrounds, and it does have useful vision to a range of about 50 feet (Gilbert, 1963).

The shark retina is poor in bipolar and ganglion cells, making shark vision of low acuity but high sensitivity in dim light. The sensitivity in dim light is enhanced by the *tapetum,* which is situated in the choroid underlying the retina and consists of a silvery plate of quanine crystals that acts as a mirror and reflects light back to the retina, which can also serve as a mirror to reflect light back to the retinal photoreceptors. Although the tapetum, when present, is normally formed in the choroid, some teleosts develop a quanine mirror in the pigment layer of the retina, which can also act as a mirror to reflect light back to the retinal photoreceptors.

Amphibious fish

The "four-eyed" blenny, *Dialommus fuscus,* of the Galapagos Islands, frequents the rocks between tides (Figure 10.5a). When I was in the Galapagos Islands, on the island of Santa Cruz, I was struck by the behavior of these amphibious fish. *Dialommus* eyes can adjust for both aerial and aquatic vision because they have two distinct optical systems. The cornea is partitioned and the lens shaped so that it refracts light onto the lower part of the retina when in air, and onto the upper part of the retina when in water. As a result, both aerial and aquatic objects are focused simultaneously on different parts of the retina. The combination of the fish's

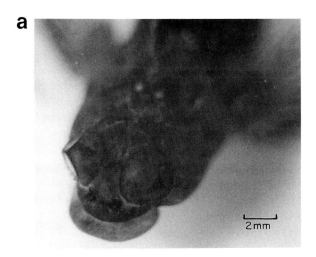

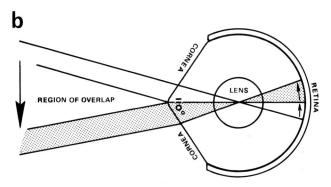

FIGURE 10.5   (a) The amphibious fish, the four-eyed blenny, *Dialommus fuscus* (Galapagos Islands). (b) Schematic of the optics of the *Dialommus* eye. (After Stevens and Parsons, 1980.)

prismatic cornea, and the index of refraction of 1.0 in air, leads to a double image of the world when out of the water (Figure 10.5b). This double view, or binocular, with a 20° to 30° overlap, may contribute to depth perception or may simply be an evolutionary "adaptive" development (Stevens and Parsons, 1980).

The mudskipper, *Periophthalmus*, is a member of the family of gobies that flourishes in mud flats along tropical shorelines from Africa through Southeast Asia to Japan. Their eyes are on retractable stalks for aerial vision in bright sunlight. Their retina is partitioned—the lower half contains primarily cones, while the upper half has only rods for vision in the mud flats.

Amphibious vision for all animals with curved corneas requires an exceptional range of focusing ability. Whereas the cornea in air focuses light like a strong positive lens, under water it becomes a very weak positive lens because it has approximately the same refractive index as sea water. Thus, land animals become hyperopic (farsighted) in water and aquatic animals become myopic (nearsighted) in air. In some fish, this gain or loss of lens power amounts to as much as 20 or 30 diopter.

Flying fish skip above the surface of water. For example, the Atlantic flying fish *Cypselurus heterus* has modified its cornea so that its vision has adapted to air and the air-water interface (Baylor, 1967). Therefore, the cornea is a pyramid (prism) instead of a curved surface or a hemisphere. The fish is then able to look up and forward through the anterior face of the pyramid, up and backward through the posterior face of the pyramid, and down through the ventral face. The flat corneal surfaces represent a means of avoiding corneal refraction (Sivak, 1980). This eye structure is likened to the surf-jumping fish described for *Dialommus fuscus*.

Deep-sea fish

The eye structure and retinal photoreceptors of fish, especially those that live in the deep sea, have been extensively studied by Munk (1966, 1980, 1984). Fish that live at depths greater than 250 meters live primarily in darkness or in very diffuse light. There is a faint blue light at these depths, from light emitted by bioluminescent organisms. Many fish also possess bioluminescent organs, or photophores, that are located around the eye and/or on various parts of the body. These fish have eyes that can detect bioluminescent flashes and thus are able to discern shapes at relatively low levels of light. The fish's visual absorption spectrum is shifted from that of land vertebrates in the green around 500 nm towards the blue around 480 nm. Thus, they have adapted their visual spectral sensitivity to their environment.

A rare deep-sea fish, *Bathylychnops exilis,* is a streamlined, javelin-shaped fish about 18 inches in length. It has two eyes (a schematic is shown in Figure 10.6). The large eye has a lens (l) and a retina (r), while the small eye has a lens (sl), retina (sg), and cornea (w). The retina has only rods and no cones. The cornea may serve to bend light into the lens of the large eye. The anatomical axis of each eye forms an angle of 35° with the perpendicular, thus providing for a large dorsal binocular field of vision. The eye is remarkable for possessing a smaller eye with a common retina that is continuous between the two eyes. The head of this dual-eyed fish is designed so that the large eye looks upward, while the small eye peers downward and backward. The fish is known as a fearsome hunter, since its extraordinary eyes offer maximum visual power in the ocean's perpetual twilight (Munk, 1966; Cohen, 1959).

Another rare deep-sea fish, *Idiacanthus,* was first described by Beebe (1934, 1935) in his descent in a bathysphere off Non Such Island Bermuda. The larval female is found between 250 and 500 meters in the sea and is from 13 to 20 cm in length (Figure 10.7a). Its two eyes, on each side of its head, are located at the ends

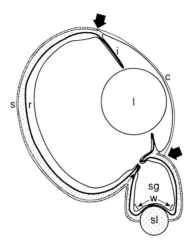

FIGURE 10.6 Cross-section of the eye of the deep-sea fish *Bathylychnops exilis;* arrows point at limbus cornea and chorid. The lumen of the secondary globe, *sg,* the scleral lens *sl,* and the window of the diverticulum retina. This fish has basically two eyes, a large eye and a small eye.

of long, thin, transparent stalks which are from one half to the full length of the body (Figure 10.7b,c). Being on long stalks enables the fish's eyes to bend and swing in all directions in order to scan its environment, a system that has been likened to a periscope. As *Idiacanthus* grows into adulthood, the stalks shorten and retract into the head.

When I became aware of this unusual fish, I began searching for *Idiacanthus* in the Bermudas. Fortunately, a larval female became available to me through the collection of Dr. J. E. Craddock, of the Woods Hole Oceanographic Institution. In examining the eye, I found that it consists of one large lens and two smaller, bulbous, lens-like structures on each side of the large lens (Figure 10.7c). Electron microscopic examination of the lens revealed it to be a layered structure, lamellae, a graded index of refraction lens (Figure 10.7d). The multiple retina consists of four distinct photoreceptor layers. There are four nerve fibers that pass from the retina down the stalks to the brain.

Deep-sea fish generally have large eyes that are oriented dorsally to increase their field of view. To improve their vision in the deep-sea environment they developed tubular or telescopic eyes (see, for example, Figure 10.8). This permits them to see silhouettes, profiles, outlining shapes, and to differentiate food from prey. Another structural feature of their retina is that the retinal rod OS are significantly longer, about 50 μm, from surface-dwelling fish, whose rod OS are much shorter, about 15 μm in length. Also, they possess a *diverteculata* that projects laterally from the eyecup through a slit-like opening to collect the re-flected light from crystals located in the *argenteum,* situated laterally near the

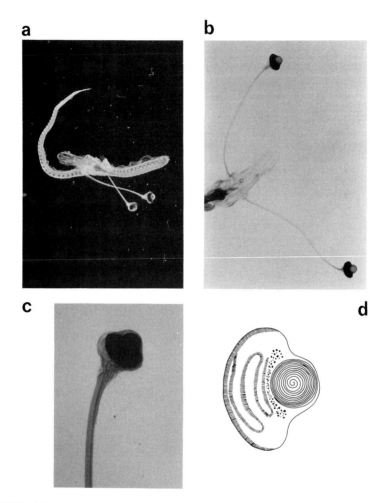

FIGURE 10.7   (a) *Idiacanthus,* deep-sea fish (larva), eyes on long stalks. (b) Enlargement of eyes showing nerve fibers that extend from the retina to the optic nerve to the brain (c). Schematic of the eye showing structure of the lens and multiple retinal layers (d).

primary lens, and this is presumed to increase their visual field (Munk, 1966, 1980).

Polarized light detection

Under water the light field is polarized due to the scattering of water molecules and particulate matter. A number of fish are known to detect polarized light (Waterman, 1984, 1989). Hawryshyn (1992) showed that some species of fish orient themselves in response to a particular angle of polarization of light. He found that

a
b

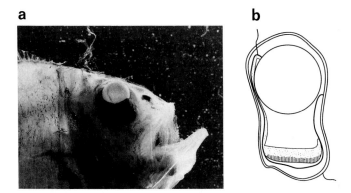

FIGURE 10.8   (a) Hatchet fish, *Argyropelecus,* tubular eye. Schematic of the eye with its large lens and retina (b).

when polarized light included the ultraviolet wavelengths, the fish was able to orient relative to the polarized light field accurately and that when ultraviolet wavelengths were absent in the stimulus, the fish did not orient to the E-vector. This would indicate that the fish's visual photoreceptors, most likely cones, are sensitive to ultraviolet light around 380 nm and that their visual pigment, rhodopsin, is aligned in a specific cone to detect the E-vector. These studies indicate that fish under water use their eyes for the detection of polarized light for orientation and navigation.

CHAPTER ELEVEN

Spectral Sensitivity
and Color Vision

VISUAL SPECTRAL SENSITIVITY

We experience the sensitivity of our eyes to light in going from bright daylight to darkness, as in entering a theater. It takes from five to ten minutes for our eyes to become dark-adapted. This phenomenon, moving from phototopic (daylight) to scototopic (night) vision, was first described by the Czech physiologist Jan Purkinje (1825) as a shift in spectral sensitivity from the blue toward the red of the visible spectrum.

The basis for this difference is that the retinal rods are sensitive at relatively low light levels and their spectral absorption is in the blue-green around 500 nm while the cones are functional in bright light (and color vision) and their spectral absorption lies more toward the red, 560 to 620 nm of the visible spectrum.

Behavioral action spectral studies to determine the sensitivity of eyes of different animal species show that their visual spectral sensitivities vary from the near ultraviolet to the infrared, from about 340 to 700 nm. This depends on the habitat of the animal and the absorption spectral peak of their visual pigments.

Evidence for this is found in the spectral sensitivities of surface fish compared to those that live at great depth. Clarke and Denton (1962) and Denton (1990) observed that the increasing blueness of light with depth in the ocean brought a shift in the absorption peak of eyes of deep-sea fish toward the blue end of the spectrum. This was confirmed by comparing the absorption spectrum of rhodopsin extracted from surface fish to that from fish living at depths of 200 to 400 meters (Muntz, 1987). The spectrum is shifted from the blue-green, around 500 nm, toward the blue, around 480 nm.

What then determines the spectral absorption maximum peak for rhodopsin? The absorption peak of retinal, the chromophore of the visual pigment, is around

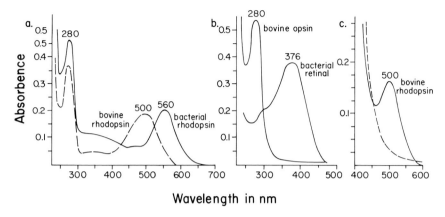

FIGURE 11.1   (a) Absorption spectra of bacteriorhodopsin (isolated from *Halobacterium halobium*) and bovine rhodopsin. (b) Absorption spectra of bacterial retinal and bovine opsin. (c) The complex of bacterial retinal and bovine opsin, forming bovine rhodopsin; after irradiation with light (----).

370 nm; but when complexed with its visual protein, opsin, to form rhodopsin, the absorption peak is shifted to longer wavelengths of around 500 nm. Retinal does not greatly influence the absorption peak for rhodopsins, and opsins do not themselves absorb light in the spectral range of from 480 to 560 nm, that of rhodopsins. But, when retinal is complexed with opsins in the eyes of different animal species, the absorption peak of the rhodopsin is determined by the protein opsin, which indicates that the visual protein opsin is species-specific.

This was experimentally determined by extracting bacteriorhodopsin from the halophilic bacterium, isolating retinal, and from frog rhodopsin, isolating the protein opsin. When retinal and opsin were complexed, the absorption spectrum was that of frog rhodopsin, and when retinal was complexed with bovine opsin, the absorption spectrum was that of bovine rhodopsin (Figure 11.1a,b,c). It was then found that the bacterial retinal can be replaced with 11-*cis* retinal (obtained from Hofman LaRoche or Eastman Kodak) and complexed either with frog or bovine opsins, the resulting absorption spectra being those of frog rhodopsin and bovine rhodopsin (Wolken, 1986; Wolken and Nakagawa, 1973). These experimental results confirm that opsin is species-specific and determines the spectral absorption peak of rhodopsin.

COLOR VISION

How do we and animals see colors? Animal eyes with only one visual photo pigment are sufficient, but to "see" with only one visual pigment, the world is monochromatic; objects appear black, white, or shades of gray. To see colors, two or more different spectrally absorbing visual pigments are needed. At least three different absorbing pigments are necessary (a trichromatic visual pigment system), but even four (a tetrachromatic visual pigment system) can be useful to distinguish all colors of the visible spectrum. The ability to discriminate various colors provides for us and animals an additional dimension in which to experience the world.

There are two general theories to account for color vision; one is the tricolor theory, which arose from the early psychophysical studies of Young (1802, 1807), von Helmholtz (1852, 1867), and Maxwell (1861, 1890). The tricolor theory asserts that there are three different visual pigments in the retinal cones, absorbing in the blue, in the green, and in the red regions of the visible spectrum. The other is the theory of Hering (1885), which postulates that there are six basic responses, which occur in pairs: blue-yellow, red-green, and black-white. Excitation leading to any single response suppresses the action of the other member of the pair. According to Hering's theory, the brain computes yellow and white from green and red at high light intensity and white from blue at low light intensity. Hurvich (1981) has reviewed these theories and psychophysical experimental studies for color vision.

How meaningful are these psychophysical theories and photobehavioral spectral sensitivity measurements related to the absorption spectra of the visual pigments in the retinal cones of the eye? Microspectrophotometry has now made it possible to obtain the absorption spectrum of an individual cone in the retina. Such absorption measurement of frog cones indicates that they have general absorption throughout the whole of the visible spectrum, with maximum around 430, 480, 540, 610, and 680 nm (Wolken, 1966). In carp retina, absorption peaks in cones were found around 420 to 430, 490 to 500, 520 to 540, 560 to 580, 620 to 640, and 670 to 680 nm (Hanaoka and Fujimoto, 1957). In goldfish, which belongs to the carp family, Marks (1963) found cones with absorption spectral peaks around 455, 530, and 624 nm, and Liebman and Entine (1964) found similar absorption peaks around 460, 540, and 640 nm (Figure 11.2). In human and monkey foveal cones, absorption spectral peaks were found at 445, 535, and 570 nm (Marks et al., 1964). In human cones, absorption spectral peaks were found around 450, 526, and 555 nm (Wald, 1964; Wald and Brown, 1965). In Figure 11.3, human and monkey cone visual pigment absorption spectra are compared to the human rod visual pigment absorption spectrum (Brown and Wald, 1964; Dowling, 1987). These cone absorption spectra maxima compare well with the psychophysical measurements of the spectral sensitivity at 430, 530, and 575 nm of the human eye and confirm that there are at least three different spectrally absorbing cone pigments for color vision: one for sensing blue, one for sensing green, and one for sensing red.

To account for these spectral absorption peaks of the visual pigments in cones, a search was made to find the genetic basis for color sensitivity, and hence color vision. This brings us back to the protein opsin, which determines the spectral absorption peaks of rhodopsins, and to the genes that code for the chemical

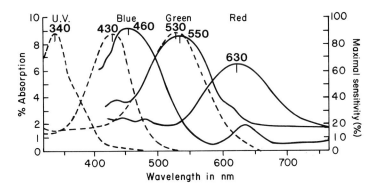

FIGURE 11.2   Absorption spectra obtained from three different cones of the goldfish (from Liebman and Entine, 1964), compared to worker honeybee, *Apis mellifera* (from Autrum and von Zwehl, 1964).

synthesis of opsins. Nathans (1992) and Nathans et al. (1986a,b) did just that in using methods of molecular genetics to isolate and identify the genes encoding opsins of rhodopsins for the green-, blue-, and red-sensitive cone visual pigments. Their experiments showed that different genes specify the synthesis of opsins for their visual rhodopsin pigments in humans, bovines, and chickens. They found that a significant homology exists between various opsins, which determines the spectral absorption peaks of rhodopsins in the retinal cones of animal eyes for color vision. In doing so, these researchers helped to clarify the genetic problems that are evident in color blindness. Recent studies of mammalian eye color vision are reviewed by Jacobs (1993).

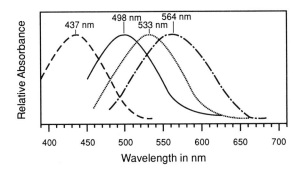

FIGURE 11.3   Difference spectra (the difference in light absorption before and after bleaching) of the three human and monkey cone pigments and human rhodopsin. The spectra were obtained by microspectrophotometry of small retinal areas. (The spectra from Brown and Wald, 1964; Dowling, 1987).

The oil globules

In the retinal cones of birds, amphibians, lizards, snakes, and turtles there are pigment oil globules (Figure 11.4). They are located in the cone ellipsoid, between the IS and the OS. Light must pass through them before being absorbed by the visual pigment in the cone's OS. These different colored oil globules in the retinal cones are in a position to act as color filters. It is of interest to inquire what function they perform in these animals.

Over a century ago, Krause (1863) suggested that these oil globules, by differentially transmitting light to the cone OS, affected the spectral response of the cone. This has led investigators to question their role in color vision. It was thought that each cone that had a colored oil globule might be most sensitive to that color and, together with the cone visual pigments, would provide sensitivity for that color and thus the basis for color discrimination.

Chicken, pigeons, and turtles necessarily function only at high light intensity, and their retinal cones predominate. In a freshly excised chicken retina, the pigmented oil globules that can be distinguished are colorless, yellow-green, orange, and red. They range in size from 3 μm to 5 μm in diameter, and their colors are beautiful to observe through a microscope.

To identify the pigment in these oil globules, three different-colored fractions were isolated from the chicken retina by Wald and Zussman (1938). These colored fractions resembled the *in situ* oil globule colors and were chemically found to be carotenoids: lutein, zeaxanthin, astaxanthin, and gallaxanthin (Wald, 1948).

The absorption spectra of the colored oil globules should indicate whether they

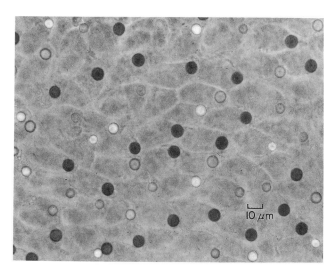

FIGURE 11.4   Pigment oil globules in the retina of the swamp turtle, *Pseudemys scripta elegans.*

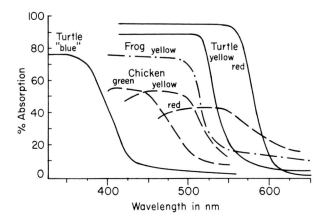

FIGURE 11.5   Absorption spectra of various colored oil globules, acting as retinal cut-off filters, in cones of the turtle (—), the chicken (----), and the frog (···).

can serve the cones for color vision. Microspectrophotometry of the *in situ* chicken oil globules (yellow-green, orange, and red) showed absorption maxima in three different regions of the spectrum (Strother and Wolken, 1960). Their absorption spectra did not have sharp peaks; they had very broad bands. The yellow-green globules had a general absorption in the region of 390 to 440 nm, the orange globules in the region of 440 to 480 nm, and the red globules in the region of 480 to 560 nm (Figure 11.5). If one assumes that the globules are acting as cut-off color filters for the cones, then the yellow-green and orange globules would seem relatively inefficient. The chicken cone pigment iodopsin has its absorption peak around 550 nm, and only the red globule has appreciable absorption in this region. The orange globule though would filter out the blue and violet light and screen out extremely bright light.

In the pigeon, colored oil globule absorption spectra maxima are around 470 to 490, 540 to 550, and 600 to 620 nm, which could account for a shift in the spectral sensitivity toward the red part of the spectrum (Strother, 1963). In comparison, Donner (1953) using electrophysiological measurements, found three "modulator" curves for the pigeon, with maxima at 470 to 490, 540 to 550, and 600 to 620 nm. On the basis of his calculated modulator curves, Donner showed that there was a shift to longer wavelengths toward the red in the spectral sensitivity of the pigeon. Therefore, these colored oil globules can account for the observed increase in spectral sensitivity to red light observed in the avian eye but not in animals without colored globules (Walls, 1942; Fox, 1953). Since a red globule absorbs light somewhere in the blue-green, the location of the globule to the cone OS could make color vision possible. In the chicken retina, Bowmaker and Knowles (1977) described five different-colored oil globules in cones and inferred that two of these colored pigmented oil globules screen for at least three types of retinal cones.

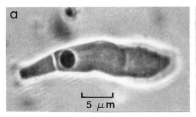
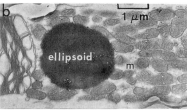

FIGURE 11.6   The frog, *Rana pipiens,* cone ellipsoid body in the inner segment and its proximity to the outer segment.

Birds have at least three different visual cone absorbing pigments and, together with their colored oil globules, could have as many as six different visual absorbing pigments that affect their spectral sensitivity, and hence their color vision (Lythgoe, 1979).

In examining the spectral sensitivity of the swamp turtle *Pseudemys elegans,* the ERG has maxima around 575, 620, and 645 nm. The absorption spectra recorded by microspectrophotometry for colored oil globules show absorption maxima for the red globules from 555 to 565 nm, for the yellow globules from 510 to 515 nm, and for the colorless globules from 370 to 380 nm, confirming the assumption that the spectral response of the turtle eye corresponds to the absorption curve of its cone pigment, cyanopsin (Strother, 1963). The spectral absorption peaks around 625 nm and near 650 nm are in agreement with two of the spectral sensitivity peaks. Absorption spectra obtained by microspectrophotometry have shown that in the turtle *Chelonia mydas* there are four different-colored oil globules: orange, two types of yellow, and colorless (Liebman and Granda, 1971). In the turtle *Pseudemys scripta,* there are red, orange, yellow, and colorless oil globules. Cones that contain either red or pale green oil globules are sensitive to red light, while cones with yellow oil globules are sensitive to green light and cones with colorless globules are sensitive to blue light (Ohtsuka, 1985). In the turtle *Emydoidea blandigii,* the yellow globule absorption spectrum matched the carotenoid zeaxanthin and that of the red and orange globules matched the carotenoid astaxanthin (Lipetz, 1984).

In amphibians, the frog retina has three types of retinal photoreceptors: "red" rods, "green" rods and cones with a yellow ellipsoid body (Figure 11.6). The green rod absorption peak is around 540 nm, the red rod absorption peak is around 610, and cones have their absorption maxima around 570 nm. These spectral absorption peaks, taken together with the yellow filtering of the ellipsoid body in cones, would cover the spectral absorption peaks for color vision.

The mammalian eye retina does not have colored oil globules, and without the color filtering effect there is no color enhancement for the cones' visual pigments. However, in primates, the fovea contains mostly cones, and the region surrounding it, the *macula lutea,* is colored yellow by the carotenoid xanthophyll (lutein). This then can act as a color filter for the three different spectral absorption peaks of visual pigments in cones. With a yellow filter, the eye has greater contrast, hence resolution, and visual acuity. In the human eye, the lens yellows with age and provides a color filter that functions like the yellow pigment globules to improve the visual acuity of aged eyes. The yellowed lens may also give the retinal photoreceptors an ultraviolet receptor. In this regard, such a light filtering system together with the retinal cone visual pigments could provide a tetrachromatic visual system instead of the trichromatic visual system by providing an ultraviolet sensitivity for daylight terrestrial animals. However, its main function may be to screen out ultraviolet light and prevent it from damaging the eye.

CONCLUDING REMARKS

The sensitivity of the vertebrate eye depends on the light intensity and the absorption spectra of the retinal rods and cones. Rods are sensitive at low light levels, and their visual pigment absorption is around 500 nm; cones are sensitive to bright light, and their visual pigment absorption is shifted toward the red, to around 560 to 620 nm of the visible spectrum.

The visual pigments are retinal-protein complexes. Specific genes encode the synthesis of color-specific forms of the protein opsin. The different forms are discretely expressed in the red cone, green cone or blue cone, thus making color vision possible. The cone visual pigment absorption spectra together with the rod absorption spectrum could provide for a tetrachromatic visual system.

In cones of birds, lizards, snakes, and amphibians there are pigmented oil globules that are located in the cone ellipsoids (between the IS and OS). These pigmented oil globules are cut-off color filters to the cone visual pigments. What other functions they serve is not precisely known. The spherical shape of these oil globules acts as a lens and together with the conical shape of the cone is a light guide, which narrows the spectral sensitivity curve of the cone and has the effect of sharpening hue discrimination. Therefore, in animals that possess pigmented oil globules, their cones act as color filters and can provide a means for processing color information.

It is of interest to note that the three different-colored filters, like those found in the pigmented oil globules, were used to develop one of the earliest color photographic processes. Louis Lumière, a French chemist, presumably without prior knowledge of the different-colored oil globules in cones, brought out an auto-chrome method in 1906 for color photography. To do so, Lumière dyed suspensions of starch grains green, blue, and red, mixed them in roughly equal proportions, and spread them over the surface of a photographic plate. The granules were squashed flat and the interstices filled with carbon black. Each colored granule served as a color filter for the photosensitive emulsion which lay beneath it, when the plate was developed, a colored photograph resulted. About the same time (around 1907), Siegfried Garten, an ophthalmologist interested in color vision, published a paper in German, based on the same principle as Lumière, on a color photographic process. This historical account of how color photographs were first processed is reviewed by Romer and Delamóir (1989).

CHAPTER TWELVE

Invertebrate Eyes: Variations in Structural Design for Vision

COMPOUND EYES

Invertebrates are the most numerous and diverse among all animal groups, and their eye structures and optical and photoreceptor systems are equally diverse. Image forming compound eyes evolved in arthropods (insects, arachnids, and crustacea) and are generally restricted to them.

Compound eyes are structured of eye facets, *ommatidia* (Figure 12.1). The number of ommatidia varies from only a few in certain species of ants to more than 2,000 in the dragonfly. Each ommatidium is a complete eye, containing two functional systems: the dioptric, or optical, system and the photoreceptor rhabdom system. The optical system is the corneal lens and the crystalline cone that transmit the image on to the rhabdom. The rhabdom is formed from retinula cells that have a differentiated photoreceptor structure, the rhabdomere. Collectively, rhabdomeres form the photosensitive *rhabdom*, or retina, within each ommatidium. The rhabdomere is analogous in function to the retinal rod OS of vertebrate eyes.

How compound eyes in insects and crustacea are able to resolve images was not known until Müller (1826) investigated the structure of insect eyes and proposed a mosaic theory of insect vision. Exner (1876, 1891), impressed by Müller's theory, set about to determine how the optical systems of compound eyes resolve images and, as a result of his experimental investigation, described two distinct types of compound eye structures: the *apposition* eye and the *superposition* eye. Exner's studies are summarized in his classical work *The Physiology of Compound Eyes in Insects and Crustacea* (1891).

According to Exner, *apposition* eyes are those in which the photoreceptor rhabdomeres that comprise the rhabdom lie directly beneath or against the crystalline cone. Each ommatidium is entirely sheathed by a double layer of pigment

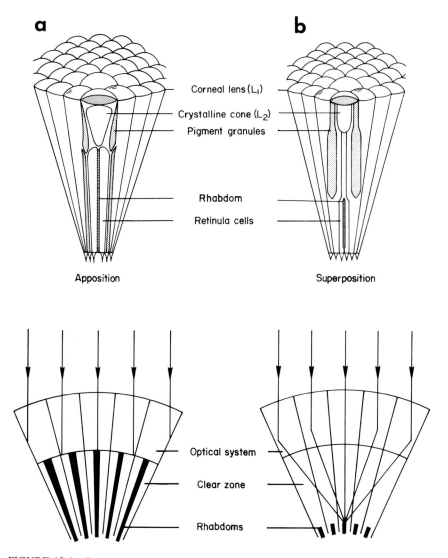

a

b

Corneal lens (L_1)

Crystalline cone (L_2)
Pigment granules

Rhabdom

Retinula cells

Apposition

Superposition

Optical system

Clear zone

Rhabdoms

FIGURE 12.1   Insect compound eye structures, eye facets (ommatidia), as schematized for (a) apposition and (b) superposition types and their optical systems.

cells. Therefore, only light striking the lens within about 10° of the perpendicular reaches the rhabdomeres. Light striking the lens at a more oblique angle may be reflected by the lens or absorbed by the pigment sheath. There can be no passage of light rays between ommatidia; the pigment sheath restricts the light rays to the rhabdomeres of that ommatidium. *Apposition* eyes are more characteristic of diurnal insects (Figure 12.1a).

Superposition eyes are those where the photoreceptor cells of the rhabdom lie some distance away from the crystalline cone (Figure 12.1b). The extent of the pigment sheath depends upon the degree of dark-adaptation of the eye. In bright light, the pigment sheath extends the full length of the ommatidium, as in the apposition eye. However, during dark-adaptation, the pigment granules migrate to the surface of the eye and are drawn up between the crystalline cones, leaving a light-permeable membrane between them. The migration of the pigment granules is analogous in function to the iris of the vertebrate eye and depends upon light intensity. At high levels of illumination, the isolation of each ommatidium is nearly perfect, each rhabdom receiving only that light which enters its own ommatidium in a nearly axial direction. At low levels of illumination, the pigment granules are retracted, allowing convergence of light from neighboring ommatidia and consequently brightening the image. Thus, light striking the surface of the eye more obliquely is not absorbed by the pigment sheath but passes through to reach the rhabdomeres. In addition, light rays from several ommatidia can be brought to focus upon the rhabdomeres of a single ommatidium, increasing the intensity of the image formed. Therefore, the superposition type eye is more efficient due to its increased light-gathering power and is characteristic of nocturnal insects. However, superposition type eyes are also found in diurnal species.

Exner (1891) also suggested that the crystalline cone of superposition eyes had lens-cylinder properties. He theorized that the greatest index of refraction was at the axis of the lens cylinder, with concentric rings of decreasing indices of refraction proceeding to the periphery of the crystalline cone, resulting in a graded index of refraction lens. A graded index lens increases the light-gathering power of the lens that would focus on the rhabdom (rhabdomeres). Exner's theory of how the optics of compound eyes form an image was way ahead of its time. The optics of compound eyes have more recently been reinvestigated by Land (1981, 1990), and Nilsson (1988, 1989a,b) and reviewed in the collected work *Facets of Vision,* edited by Stavenga and Hardie (1989).

INSECT EYES

Insects are found everywhere throughout the world. Many are diurnal, others are nocturnal, some are predatory, and all are highly specialized. How insect eyes visualize their world has been, and continues to be, of interest in visual science since the early studies of Exner. Among the varied diurnal and nocturnal insects that we have studied, only a few were selected here to illustrate how their compound eyes are structured for vision.

Drosophila melanogaster

The common fruitfly *Drosophila melanogaster* is an example of a diurnal insect. Its compound eye is composed of over 700 ommatidia, approximately 10 μm in

a   b

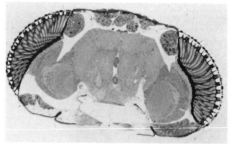

FIGURE 12.2   (a) The compound eye of the insect *Drosophila melanogaster.* (b) Cross-section through the head of *Drosophila,* showing the eye facets (ommatidia) of the two eyes. (Courtesy of Professor John Pollock.)

diameter and from 70 to 125 μm in length (Figure 12.2). Each ommatidium consists of a corneal lens, a crystalline cone, retinula cells, and a sheath of pigment cells that extends the entire length of the rhabdom (Figure 12.3). In cross-section, there are eight radially arranged retinula cells that form the rhabdom. Each retinula cell has a medial portion extended toward the center of the ommatidium and terminating in a dense circular rhabdomere. Rhabdomeres are a continuous membrane elaboration of the retinula cell surface.

The rhabdom consists of eight individual rhabdomeres (R_1 to R_8) but only seven are observed in Figure 12.3b,c. They are situated in a relatively clear fluid cavity or inter-retinular space. The individual rhabdomeres average 1.2 μm in diameter and are more than 60 μm in length. The rhabdomere fine-structure depends on the angle of the cut through the rhabdomere. In all cross-sections of the rhabdomere, the lamellar structure is observed, while in all oblique sections a microtubular structure is found (Figure 12.3d,e). A single structure of microtubules can produce both of these geometries in a thin section, depending upon the orientation of the individual rhabdomere with respect to the plane of cutting. The rhabdomere is then structured of microvilli that form rod-like microtubules, each being about 500 Å in diameter, whose double membrane is of the order 100 Å in thickness (Wolken, 1975). It is in these microtubular membranes that the visual protein rhodopsin, is molecularly associated. Only the wavelengths of light absorbed by the visual photoreceptors, rhabdomeres in the rhabdom (Figure 12.4) are transmitted through the corneal lens.

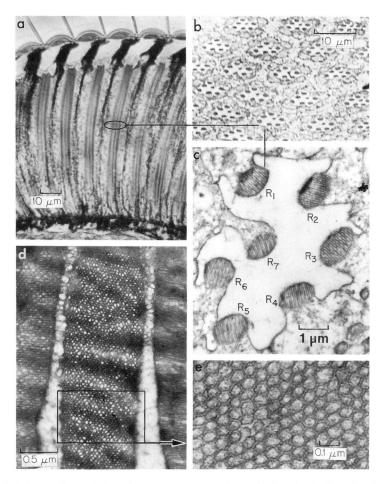

FIGURE 12.3　*Drosophila melanogaster* compound eye. (a) Longitudinal section through several ommatidia showing the corneal lens, crystalline cone, rhabdom, and pigment sheath; (b) cross-section through ommatidia; (c) cross-section through the rhabdom to illustrate the orientation of the rhabdomeres (R_1–R_7); (d) longitudinal section at the distal end of the ommatidium showing three adjacent rhabdomeres; (e) enlarged area of the rhabdomere showing the structure of microtubules.

The cockroach eye

A nocturnal insect is the cockroach, which is believed to be one of the more primitive of the unspecialized insects. The compound eye structures of two large species of cockroach, *Periplaneta americana* and *Blaberus giganteus,* were investigated (Wolken and Gupta, 1961). Their compound eyes are structured of approx-

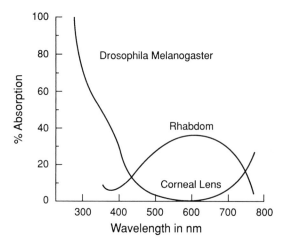

FIGURE 12.4 Light transmission through the corneal lens (transparent to wavelengths from about 450 to beyond 700 nm); absorption spectrum of the rhabdom (which absorbs from about 450 to beyond 700 nm). The light transmitted is of the wavelengths absorbed by the visual pigments in the rhabdom.

imately 2,000 ommatidia. Ommatidia are sheathed from one another by pigment granules. An ommatidium consists of seven retinula cells. Each retinula cell is from 7 to 9 μm in diameter with a large elliptical nucleus. Although only seven retinula cells were observed in *Periplaneta,* eight retinula cells were found in the cockroaches *Blatta (Stylopyga) orientalis* and *Blatella germanica* (Jorschke, 1914; Nowikoff, 1932). The eighth cell is probably a rudimentary structure located close to the basal membrane. It does not extend the entire length of the rhabdom, and no rhabdomere is differentiated from it. Aggregates of intracellular pigment granules, which do not seem to be affected by dark-adaptation, surround the rhabdoms and extend the whole length of the retinula cells. Depending upon the location and angle of cut of the eye section, the rhabdom appears either rhomboidal or triangular (with sides measuring from 5 to 12 μm in length). Each rhabdom is made up of four rhabdomeres (R_1 to R_4) which lie in close proximity and form a regular pattern of organization (Figure 12.5). A rhabdomere averages 2 μm in diameter and about 100 μm in length and is structured of microtubules, as in the *Drosophila* compound eye. These microtubules are about 500 Å in diameter, with outer bimolecular membranes of the order of 100 Å in thickness. There are approximately 400 microtubules in one square micrometer of surface, or about 80,000 microtubules in a single rhabdomere (Wolken and Gupta, 1961). The closed-type rhabdom observed in the cockroach is most likely an efficiency mechanism for light capture by nocturnal insects, for the cross-sectional area of the rhabdom in the cockroach is about five times that of the *Drosophila* rhabdom.

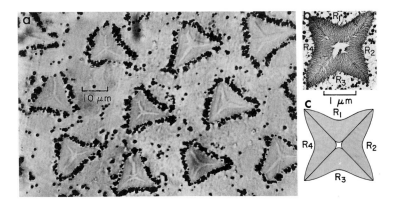

FIGURE 12.5   The cockroach *Periplaneta americana*. (a) Cross-section through many ommatidia, (b) rhabdom formed of four rhabdomeres (R_1–R_4), and (c) diagram of fused rhabdom.

The firefly eye

Among the nocturnal insects are fireflies. They are found on every continent except Antarctica, and of the more than two thousand known species of firefly, about one hundred inhabit the United States. The two most common species are *Photuris pennsylvanica,* and the North American *Photuris pyralis*. Fireflies have attracted the curiosity of humans for centuries due to their bioluminescent flashes of blue-green light in darkness.

Exner (1891) used the firefly as a model to derive his theory of the superposition type compound eye. The firefly compound eye provides an example of an ommatidium which is structurally unique among insect eyes. More recently the structure of the compound eye of the firefly *Photuris* was studied (Horridge, 1968, 1969, 1975; Wolken, 1971, 1975). The *Photuris pennsylvanica* compound eye consists of several hundred ommatidia; the ommatidium differs from previously described insect eyes in that the corneal lens extends into the region normally occupied by the crystalline cone (Figure 12.6a). A unique structural feature in the firefly eye is a crystalline thread which extends from the cone to the rhabdom. This thread is a fiber optic light guide, providing light directly to the rhabdomeres in the rhabdom (Figure 12.6a). Only the light contained in the crystalline thread is effective for stimulating the retinula cells. A similar fiber optic structure occurs in the worker honeybee, where the closed rhabdom and the surrounding zone act together as a wave guide (Varela and Wiitanen, 1970).

The corneal lens in cross-section is seen to form spirals, and in a longitudinal section these appear as laminations (Figure 12.6b–d). The corneal lens is sur-

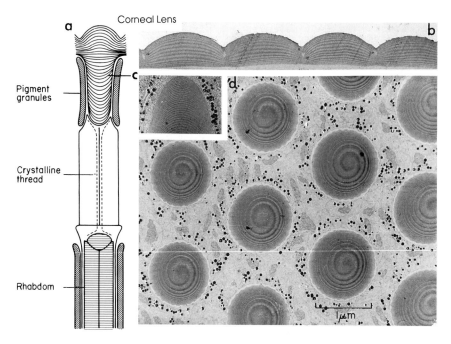

FIGURE 12.6   An ommatidium of the firefly, *Photuris pennsylvanica,* showing how it is structured (a), corneal lenses (b), and electron micrographs (c, d) of a cross-section of corneal lenses. (From Wolken, 1971, 1975.)

rounded by many distal pigment cells. Four cone cells which extend to the basement membrane as fine filaments similar to cone cell extensions are found in other insects. Their nuclei are situated below the corneal lens (Goldsmith, 1962; Horridge, 1966; Perralet and Baumann, 1969). These filaments or cone cell threads are believed to be fiber optic elements functioning as wave guides directing light to the photoreceptor (Horridge, 1968, 1986; Døving and Miller, 1969).

The photoreceptor system of retinula cells that forms the rhabdom in the firefly occupies only one-fourth of the ommatidial length. The system has two nuclear layers: a distal layer of retinula cell nuclei and a basal layer of nuclei. For each ommatidium there appears to be a single basal cell which surrounds the axonal terminals of the retinula cells. These basal cells could serve to insulate the retinula cell axons in the manner of a Schwann cell. The ommatidium contains eight retinula cells that give rise to the six rhabdomeres forming the rhabdom (Figure 12.7). Each arm has a V-shaped rhabdomere formed from the rhabdomeres of adjacent retinula cells. The small distal rhabdomere is formed by two rhabdomeres from two retinula cells. The rhabdomere fine structure is formed of microvilli into microtubules.

Firefly eyes are well adapted for light-gathering. The major portion of the

FIGURE 12.7   The rhabdom of the firefly, *Photuris pennsylvanica*, enlarged to show geometric arrangement of its rhabdomeres (R_1–R_6).

rhabdom occupies nearly the entire cross-sectional area of the ommatidium and is tightly packed against the rhabdom of neighboring ommatidia. In general, nocturnal insects such as cockroaches and moths (Fernandez-Morán, 1958; Wolken and Gupta, 1961) have rhabdoms that occupy a large portion of the ommatidium, while diurnal insects have relatively small rhabdoms (Fernandez-Morán, 1958; Goldsmith, 1962; Wolken, 1968b).

We can now compare the compound eye structure of the fruitfly, cockroach, and firefly to the rhabdom structure of some other insects. The housefly, *Musca domestica,* has an open-type rhabdom of seven rhabdomeres (Fernandez-Morán, 1958; Eichenbaum and Goldsmith, 1968) similar to *Drosophila* (Figure 12.3). In the baldface hornet, *Vespa maculata,* the compound eye ommatidium contains eight retinula cells. There is a ninth retinula cell located more distally than the other eight in a few ommatidia. The rhabdom is composed of four pairs of rhabdomeres. The crystalline cone is built of four cellular segments which taper into long tubular cone cell extensions (this is also found for the clothes moth), ending just distal to the basement membrane. There are no connections between these cone cell extensions and the nerve cells. On the other hand, the cone cell extensions have a definite structural relationship to the pigment cells which appear to make synaptic connections with them.

The honeybee, *Apis mellifera,* has several thousand ommatidia in its compound eyes. The rhabdom of the worker bee consists of four rhabdomeres, although it is formed from eight retinula cells (Goldsmith, 1962). The rhabdom structure is similar to that of the hornet, and each rhabdomere is a closely packed parallel array of microtubules with axes perpendicular to the axis of the rhabdom. The microtubules in adjacent rhabdomeres of the rhabdom are mutually perpendicular.

The compound eye of the carpenter ant, *Camponotus herculenus,* appears to represent a transition from the apposition-type eye (Figure 12.1a) to that of the pseudo-type eye with a crystalline thread, as seen in the firefly and the hornet (Figure 12.6). The retinula cells of the carpenter ant, like those of the hornet, show two zones: a clear zone near the rhabdom and a peripheral cytoplasmic zone. The rhabdom of the carpenter ant is circular (Figure 12.8) and occupies a much larger portion of the ommatidial cross-section than that in the hornet. Despite its size, the rhabdom is formed from only six retinula cells as compared with seven to nine in other insects (Vowles, 1954).

The hammerhead fly, *Cyrtodiopsis whitei* (Dipsodae among the diptera), is found in Africa and Malaysia. *Cyrtodiopsis'* compound eyes are not located in the head as in most insects, but on long stalks that extend to more than half of the body length (in males about 10 mm or more). The eyes being on long stalks permits the insect a greater field of view when scanning its environment (Figure 12.9). This is due to the fact that their binocular field comprises about 70 percent of their compound eye ommatidia, with an interommatidial angle of about 1°. Therefore, the hammerhead fly is better able to perceive its environment, to measure the distance of an object, and to gauge the size of nearby objects. This provides *Cyrtodiopsis* a great advantage over other predatory insects (Burkhardt and De la Motte, 1983, 1985; Schwind, 1989).

Other unusual eyes are found among arachnids; for example, the jumping spider, *Portia,* has two principal eyes, two anterior eyes, and two posterior eyes (Figure 12.10). Therefore, the jumping spider eyes provide a considerable field of view.

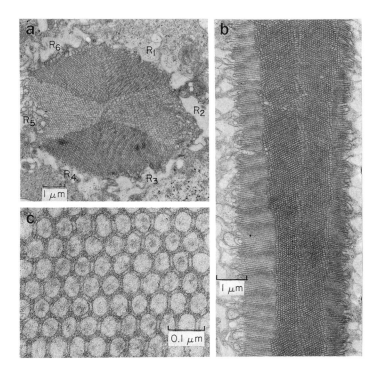

FIGURE 12.8   The carpenter ant, *Camponotus herculenus*. (a) Cross-section through the rhabdom showing rhabdomeres (R_1–R_6). (b) Longitudinal section of the rhabdom. (c) Higher resolution of a cross-section through the rhabdomere showing that it is structured of microtubules. Electron micrographs.

FIGURE 12.9   The stalked eye fly, Diopsidae, *Cytodiopsis whitei*. (From J. E. Rawlins, Carnegie Museum of Natural History, Pittsburgh, PA.)

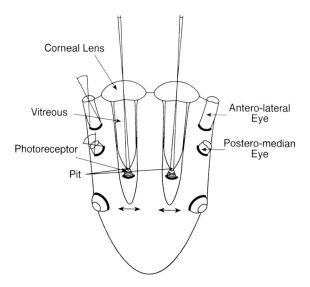

FIGURE 12.10   Diagrammatic cross-section through the head of the jumping spider, *Portia,* showing principal eyes and anterior and posterior eyes, indicating their relative sizes. (Schematized from Land, 1985.)

Tracheoles

In June beetles, butterflies, moths, and other insects, structures that surround the rhabdom are tracheoles. They are chemically formed of the polysaccharide polymer chitin that is structured into a twisted lamellar pattern—like a screw (Figures 12.11, 12.12). According to Miller and Bernard (1968), tracheoles are responsible for the eye glow of moths and the red glow of butterflies, due to the fact that the screw-like structure can function as a quarter-wavelength interference filter. The thickness and spacing of the lamellae (ridges in a screw) indicate a correlation with the reflected color of the eye glow (Brown and Wolken, 1979). Tracheoles are also found in the insect cuticle where they are associated with respiratory mechanisms of oxygen transport photophores. In the firefly lantern tracheoles, oxygen is necessary for the luciferin-luciferase bioluminescent system, resulting in the observed blue-green flashes seen for fireflies.

The corneal lens structure

The corneal lens of insect and crustacean eyes is derivable from the animal's skin, or cuticle. The cuticle is chemically formed of the polysaccharide polymer chitin. The chemical structure and optical properties of insect cuticles were studied by Neville and Caveney (1969) and Neville (1975). They found that the cuticle structure is a fibrous, layered structure and that the orientation of the molecules in

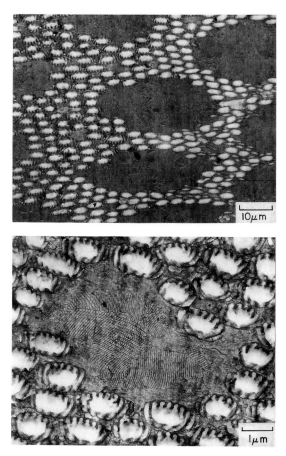

FIGURE 12.11   June beetle, *Scarab phyllophaga*, cross-section through compound eye in photoreceptor area showing rhabdoms surrounded by tracheoles. (From Wolken, 1971; Brown and Wolken, 1979.)

these layers is like that of a cholesteric nematic liquid crystal (Brown and Wolken, 1979). In cholesteric nematic liquid crystals, the molecules pack in layers and maintain a parallel orientation to each other. Thus, the direction of the long axis of the molecules in a chosen layer is slightly displaced from the direction of the axis in adjacent layers and produces a helical structure, as indicated in Figure 12.13. Bouligand (1972) has drawn a comparison between the cuticle cholesteric nematic liquid crystal structure and the structure of the corneal lens.

Electron microscopy of the corneal lens of the firefly (*Photuris pennsylvanica*) and the June beetle (*Scarab phyllophaga*) shows that they are structured of concentric lamellae. In each successive lamellar layer, the direction is rotated about the axis perpendicular to the planes. That is, the direction of the layers rotates through

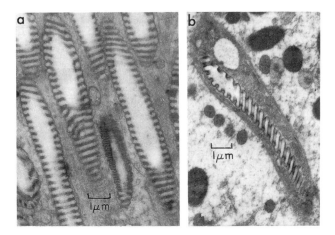

FIGURE 12.12a,b   Trachcole structure, longitudinal sections in (a) June beetle, *Scarab phyllophaga,* in rhabdom photoreceptor area; (b) firefly, *Photuris pennsylvanica,* in photocyte cells of the lantern. (From Wolken, 1975.)

180° from one layer to the next, forming a helix. The arrangement of layers is not in a series of concentric ellipsoids as Exner (1891) proposed but in a series of paraboloids along the radial axis (Figure 12.13b, 12.14a–c). The corneal lenses of the carpenter ant (*Camponotus herculenus*), the fruitfly (*Drosophila melanogaster*), and the housefly (*Musca domestica*) show a similar structural arrangement, which is probably found in other insects as well (Wolken, 1986).

   Cajal (1918) on tracing out the visual system of vertebrates, thought that the

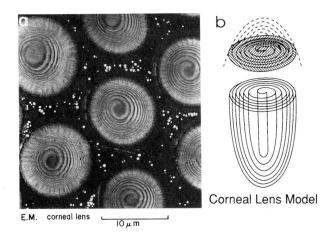

FIGURE 12.13   (a) Corneal lens structure of the June beetle, *Scarab phyllophaga.* (b) A schematized structural model of the corneal lens.

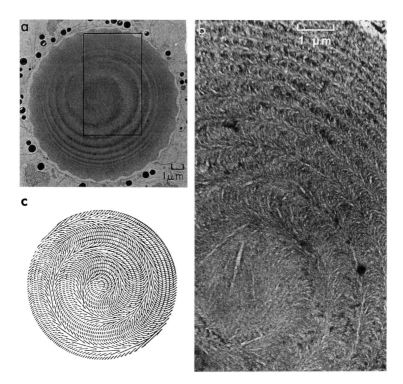

FIGURE 12.14   (a) Corneal lens of firefly, *Photuris pennsylvanica.* (b) Enlarged section of (a). (c) Diagrammatic representation, projection on the cutting plane of the fibril directions in the concentric rings.

general plan for all visual systems would be found in the insect eye, but after studying the insect eye, he wrote: "The complexity of the nerve structure (the optic lobe) for vision is even in the insect something incredibly stupendous—a marvel of detail and precision."

CRUSTACEA

A variety of freshwater and marine species are found among the crustacea—from lobsters and crabs, which are considered giants, to the tiny water fleas, which are only a few millimeters long. Most species, like insects, possess simple eyes and compound eyes, the exception being copepods. It is of interest to examine several freshwater and marine crustacea whose eyes exhibit some general structural as well as special features in their optical and photoreceptor systems.

The freshwater *Daphnia,* a water flea, possesses a compound eye and a simple "nauplius" eye (Wolken, 1971). The compound eye of *Daphnia pullex* consists of

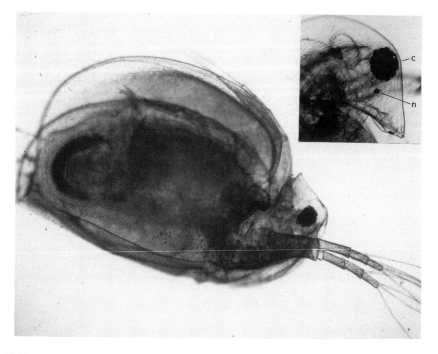

FIGURE 12.15   The waterflea, *Daphnia pulex*. (c) compound eye; and n, naplius eye.
Light micrographs.

about twenty-two ommatidia enclosed in a capsule under the cuticle of the animal,
which serves as a common lens for all the ommatidia (Figure 12.15). The eye is in
continuous oscillatory motion. The corneal lens and the distal pigment cells in the
insect ommatidium are not usually found. The ommatidium has a crystalline cone
and elongated retinula cells that are surrounded by pigment granules (Figure
12.16). There are eight retinula cells that give rise to seven rhabdomeres that form
a closed-type rhabdom (Röhlich and Törö, 1961).

The rhabdomeres that form the rhabdom consist of microtubules (microvilli)
like the insect rhabdomeres. The microtubules within each rhabdomere are precisely arranged with their longitudinal axes regularly aligned in a given direction
for one set of rhabdomeres and in the perpendicular direction for alternate layers.
Such an arrangement of the rhabdomeres could be a structural basis for the ability
of *Daphnia* and other crustacea, such as the land crab *Cardisoma* and the swimming crab *Callinectes* (Eguchi and Waterman, 1966), to analyze the direction of
polarized light (Waterman et al., 1969; Baylor and Smith, 1953).

Another freshwater crustacean is the carnivorous water flea *Leptodora kindtii*,
which is relatively large (18 mm in length) in comparison with *Daphnia*. *Leptodora* is almost completely transparent, with one median spherical compound eye

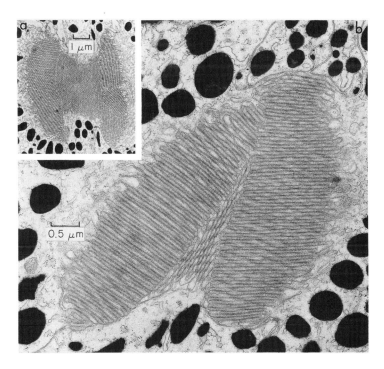

FIGURE 12.16   The waterflea, *Daphnia pulex*. Rhabdom (a) and enlarged rhabdomeres (b) show microstructure of microvilli.

(Figure 12.17). The eye of *Leptodora kindtii* is located at the anterior end of the organism and is contained entirely within the transparent exoskeleton that functions as a corneal lens. The eye is free to move and can rotate 10° in either direction to scan its environment (Wolken, 1975). A small area behind the eye with connecting neurons accommodates the optic process leading to the brain (Scharrer, 1964). *Leptodora* is believed to have better vision than most crustacea, for it can rapidly capture copepods as large and as fast as Cyclops.

The compound eye of *Leptodora* is composed of about 500 ommatidia radially arranged. The ommatidia are large conical structures, measuring about 180 μm in length with a diameter of 30 μm at the outer portion and 2 μm at the base. A schematic ommatidium and a cross-section through the crystalline cone and rhabdom are seen in Figure 12.17a–e. The crystalline cone constitutes about two-thirds of the ommatidial length. Although it is rounded at the outer end, there is no evidence of a distinguishable lens cap. There is, however, a completely transparent interstitial space between the surface of the eye sphere and the external chitinous wall, which functions as a common lens for all the ommatidia. The crystalline cone is composed of five equal pie-shaped segments, formed from five crystalline

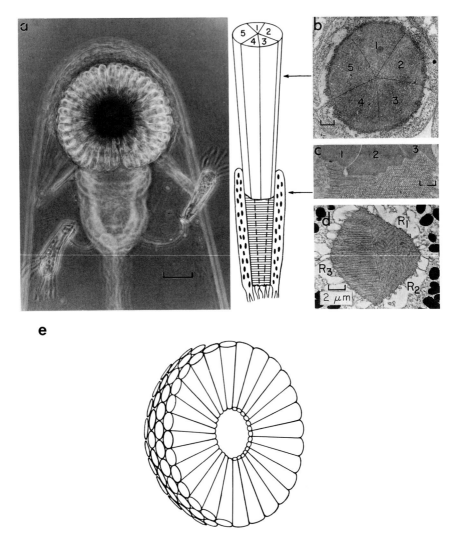

FIGURE 12.17   The waterflea, *Leptodora kindtii*. (a) The eye. The animal's cuticle acts as a common lens. (b) Cross-section of crystalline cone composed of five wedge-shaped segments (1–5). (c) Longitudinal section showing the connection of the crystalline cone with the rhabdom. (d) Cross-section of the rhabdom indicating rhabdomeres R_1–R_3 and their orientation. (e) Diagrammatic model of *Leptodora* eye.

cone cells which serve to concentrate the light into a narrow beam (Figure 12.17b). As the crystalline cones continue inward, the space between them increases and is filled with pigment cells.

It is not known whether the segments of the crystalline cones promote a system of total internal reflection. Although structural observations show that the crystalline cone continues proximal to the surface of the rhabdom as in the apposition-type eye, observations of pigment migration indicate that under certain conditions of dark-adaptation, "crossing" among adjacent crystalline cones could result in the formation of a superposition image. The rhabdom is affixed directly to the ends of the crystalline cones. The four radially arranged retinula cells that form the rhabdom show only three rhabdomeres. Cross-sections through numerous ommatidia in the rhabdom area reveal the rhabdom structure. One of the rhabdomeres, R_3, is large in comparison with the other two, R_1 and R_2, and appears to be comprised of two smaller rhabdomeres that have fused (Wolken and Gallik, 1965). Four retinula cells yielding only three distinguishable rhabdomeres for the rhabdom have also been observed for the dragonfly (Goldsmith and Philpott, 1957; Naka, 1960).

The rhabdomere *fine structure* of *Leptodora* is that of tightly-packed, microtubular microvilli. The microtubules of the small rhabdomeres R_1 and R_2 are arranged perpendicularly to those of the large rhabdomere R_3. The ends of the microtubules appear continuous with the cytoplasm (Figure 12.18b). This has also been noted for the rhabdomeres of many other crustacea, and this structural connection with the cytoplasm may be of some importance to visual excitation (Lassansky, 1967).

It is of interest to point out that the corneal lenses, as in all insects, are hexagonally packed structures, while among the crustacea (shrimp, crayfish, and lobster) they are squares (Figure 12.19). For example, the lobster eye corneal lenses are radially oriented squares. According to Vogt (1977) and Land (1980), light entering the crystalline cone (or rod pyramid) is reflected from one or more of its sides and is focused on the photoreceptors in the rhabdom. Accordingly, it is highly probable that these square corneal lenses function optically as mirrors, making them analogous to a reflecting superposition eye (Figure 12.19b).

Copepods

One of the more interesting and unique eyes found among crustaceans is the relatively rare marine copepod, *Copilia quadrata*. The eyes of this crustacean have been of considerable curiosity since the animal was first found in the Bay of Naples, Italy, and described by Grenacher (1879). Exner (1891), working in Naples, rediscovered *Copilia* and was fascinated by its strange pair of eyes which he likened to a telescope with two lenses. The "eyepiece lens" is deep in the animal's body and is in continual motion (Figures 12.20, 12.21).

Copilia quadrata is about 1 mm wide and 3 mm long. Only the female of the species possesses these scanning eyes, for the male is blind. The *Copilia* eye structure has more recently been described by Vaissiere (1961), Gregory (1966, 1967), Wolken (1975), and Wolken and Florida (1969). The *Copilia* eye resembles an ommatidium of the compound eye with a biconvex corneal lens (anterior biconvex lens, L_1) and, at some distance away, a novel "pear-shaped" crystalline cone (posterior lens, L_2). Attached to the crystalline cone are the retinula cells, which give rise to rhabdomeres that form the rhabdom in the retina. The rhabdom

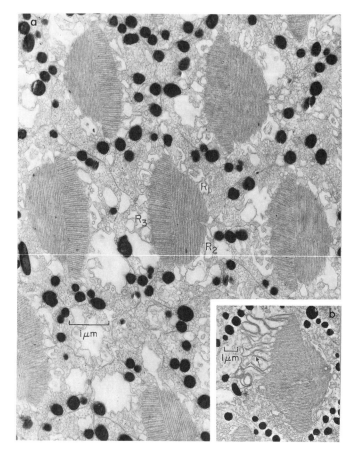

FIGURE 12.18   The waterflea, *Leptodora kindtii*. (a) Cross-section through many ommatidia and their rhabdoms, and (b) rhabdom, cross-section at higher magnification (arrow) showing the relationship of the rhabdomere (microvilli) with the cytoplasm.

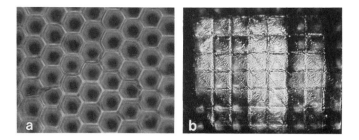

FIGURE 12.19   The corneal lenses of (a) insect compound eye corneal lenses compared to (b) the lobster eye corneal lenses.

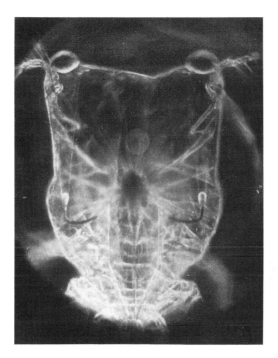

FIGURE 12.20   The copepod, *Copilia quadrata, Mediterranean,* darkfield photomicrograph. (Courtesy of Dr. Neville Moray, York University, Toronto, Canada.)

lies in an L-shaped, orange-colored stem that oscillates back and forth in a sawtoothed pattern, varying from about one scan to five scans every one to two seconds (Gregory 1966, 1967; Moray, 1972). The stems from both eyes move synchronously and rapidly toward each other then separate slowly. Gregory (1966) has likened such scanning to a television camera: "It seems that the pattern of dark and light of the image is not given simultaneously by many receptors, as in other eyes, but in a time-series down the optic nerve, as in the single channel of a television camera."

In *Copilia quadrata,* the retinula cells lie directly behind the crystalline cone, followed by the rhabdomeres that comprise the rhabdom (Figure 12.22). The rhabdom measures 11×17 μm and is completely surrounded by pigment granules. It extends 60 μm in length from the retinula cells to the bend of the stem. Only five rhabdomeres (R_1–R_5) can be identified in the rhabdom. One of them, R_1, is an asymmetric rhabdomere, located at a nodule on the side of the stem facing the brain and lying at the base of the crystalline cone. The asymmetric rhabdomere appears to lie at nearly 45° with respect to the stem. Rhabdomeres R_2–R_5 are ellipsoids measuring 1×2 μm in diameter and about 58 μm in length. These rhabdomeres lie with their longest dimension parallel to the stem. Mito-

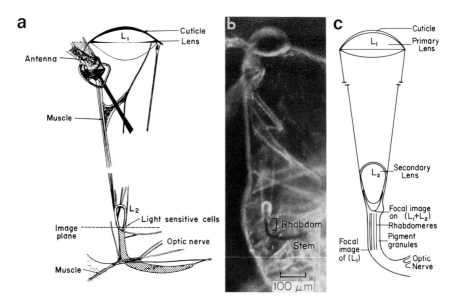

FIGURE 12.21   (a) The structure of the *Copilia* eye, as drawn by Grenacher (1879). (b) *Copilia quadrata,* darkfield photomicrograph of the eye. (c) The optical system showing the positions of the corneal anterior lens (L_1) and the crystalline cone, posterior lens (L_2); focal point of the corneal lens L_1; focal point of L_1 and L_2 in the total optical system. (From Wolken, 1975, p. 159.)

chondria and structures resembling synaptic vesicles are also found in this region. Rhabdomeres R_1 to R_3 are separated by screening pigment granules, whereas rhabdomeres R_4 and R_5 are not. The rhabdomere *fine structure* is that of microvilli forming into microtubules, whose structure is similar to that described for the insect rhabdomeres.

   Copilia lives at depths greater than 300 meters, where light comes primarily from bioluminescent organisms and fish photophores. The level of light is relatively low, and it was expected that the *Copilia* rhabdoms would be of the closed-type, providing a more effective cross-section for light-gathering; but the rhabdom is of the open-type, common in most diptera that navigate at high light levels (Figure 12.22).

   The *Copilia* eye can be considered analogous to the superposition-type ommatidium in which the crystalline cone lies at some distance from the corneal lens. In addition, the crystalline cone forms a convex interface with a fluid of lower refractive index. The concentration of this material varies across the diameter of the crystalline cone (L_2), the greatest concentration being in the center, indicating that the lens has a graded index of refraction (a GRIN lens). The crystalline cone L_2 can function as a lens, and the shape of the lens is optimized for collecting extremely weak, diffuse light.

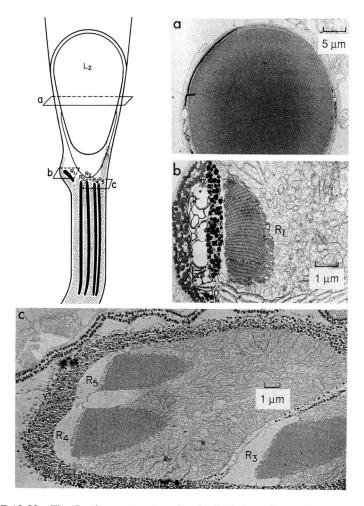

FIGURE 12.22   The *Copilia* eye structure, longitudinal view of crystalline cone and rhabdom. Electron micrographs. (a) Crystalline cone, cross-section (L_2); (b) oblique section showing asymmetric rhabdomere (R_1) in rhabdom; (c) oblique section through the rhabdom showing three rhabdomeres (R_3–R_5) of the five rhabdomeres that form the rhabdom.

The *Copilia* eye with its corneal lens, L_1, and its crystalline cone lens, L_2, is a two-lens optical system in which the lens L_2 is positioned a short distance in front of the rhabdom and focuses the image on the rhabdom (Wolken and Florida, 1969; Wolken, 1975, 1986). The optical system is such that the corneal lens, L_1, forms an image at I, which is intercepted by the posterior lens, L_2, and imaged at I_2 (Figure 12.23). The effect of the second lens, L_2, is to condense the partially focused image from L_1 onto a much smaller area, thereby increasing brightness and acting as a light "amplifier." The nodal point (N) of the combination is found

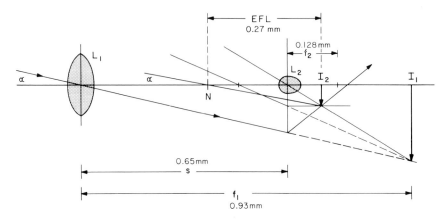

FIGURE 12.23   Optical system of the *Copilia* eye.

by drawing a line parallel to the original ray and passing through the final image; this line behaves like a ray passing through a single lens at N. The effective focal length (EFL) is the distance from N to I_2, and is obtained from the equation:

$$EFL = \frac{f_1 \times f_2}{f_1 + f_2 - s}$$

where f_1 and f_2 are the focal lengths (in water) of the anterior lens, L_1, and the posterior lens, L_2, and s is the separation of their centers. From our data on the *Copilia* eye, $f_1 = 0.93$ mm, $f_2 = 0.128$ mm, and s = 0.65 mm. Thus, the EFL is 0.27 mm. This value brings the f-number down to 1.6, which is comparable to that of a fish eye lens. The importance of this optical system is that it greatly facilitates scanning and provides for a high-aperture, high-resolution optical device. *Copilia* then has adapted to the very low light levels by evolving a remarkably advanced optical system that maximizes the collection of very diffuse light in its environment. How such a lens and optical system may be adapted as an aid for the visually impaired and further technological applications are discussed in Chapter 15.

An equally rare copepod that lives well offshore on the water surface is *Pontella spinipes*. According to Land (1980), their eyes may be just as remarkable as the *Copilia* eye. In the *Pontella* male eye, the optical system has triple, instead of dual, lenses. Land (1980) assumed a uniform index of refraction of 1.53 for the lenses and showed how such a triple lens would focus and image with its photoreceptors, the rhabdom that lies immediately behind the third lens as indicated in Figure 12.24.

The horseshoe crab, *Limulus polyphemus* (Figure 12.25), is not a crab, for it belongs to the class Merostomata, that is closely related to arachnids and a distant relative of crustacea. *Limulus* has a long history, going back 350 million years; it is a living fossil that continues to inhabit the shallow waters along the Atlantic coast.

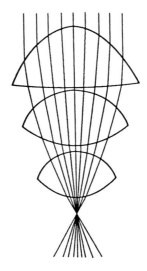

FIGURE 12.24   Three lens optical systems of the eye of the male *Pontella spinipes*. (According to Land, 1981; Pumphrey, 1961.)

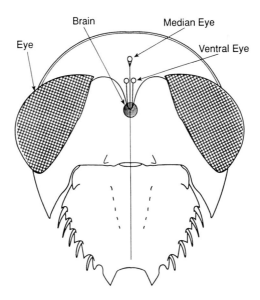

FIGURE 12.25   The horseshoe crab, *Limulus* eyes, schematic drawing.

The evolutionary history and behavior of *Limulus* has been a curiosity to naturalists who have observed and studied its breeding cycle for a very long time. The visual system of *Limulus* was a life-long study of H. K. Hartline with F. Ratliff and associates at Rockefeller University. Their studies are summarized in the collected papers (Hartline, 1974). Hartline was awarded the Nobel Prize for vision in 1967. Barlow (1990) and his colleagues have extended these studies of the *Limulus* behavior to studies of the visual system and the relationship between the eye and brain.

On examining the *Limulus* lateral compound eye, there are about 800 ommatidia, covered by a transparent chitinous membrane that acts as the corneal lens for all the ommatidia. The *Limulus* eye has no useful refracting surface in water, and the question arose whether *Limulus* has an imaging eye. In order to form an image, the lens acts as a lens-cylinder, with the highest refractive index along its axis, decreasing parabolically in the index of refraction for increased light-gathering (as depicted for corneal lenses of the June beetle, Figure 12.13, and the firefly, Figure 12.14). Behind the lens is the rhabdom that is formed from eight to twelve rhabdomeres around an eccentric cell. The rhabdomeres that form the rhabdom are arrayed like spokes of a wheel.

TRILOBITES

> *The eye of the trilobite tells us that the sun shone on the old beach where he lived; for there is nothing in nature without a purpose, and when so complicated an organ was made to receive the light, there must have been light to enter it.*
> —JEAN LOUIS RODOLPHE AGASSIZ, 1870, *Geological Sketches*

Trilobites developed in the Paleozoic era and flourished in the Cambrian 500–600 million years ago, only to become extinct about 250 million years ago. Species of trilobite fossils have been uncovered in various parts of the world. Their compound eyes appear essentially similar to present-day arthropod compound eyes. Their eyes have been a curiosity to scientists for some time, for they are believed to be the earliest animal to have evolved compound eyes.

The trilobite compound eyes appear essentially the same as present day arthropod compound eyes. Stuermer (1970) and Towe (1973) investigated the structure of *Phacops* corneal lenses by electron microscopy and x-ray analysis. Their studies revealed that the corneal lenses were mineralized calcite (crystallized $CaCO_3$) that is highly oriented and behaves optically like glass. Towe (1973) indicated that calcite was most likely present in the living trilobite and that such a lens is capable of forming an inverted image over a large depth of field.

The corneal lenses of the *Phacops rana* are arranged in vertical rows and appear similar to present-day arthropod compound eyes (Figure 12.26). In his studies, Stuermer (1970) found filamentary fibers leading from the corneal lens to the

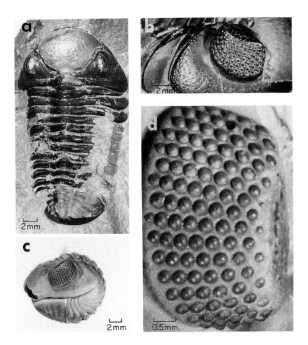

FIGURE 12.26   (a) Trilobite, *Phacops rana,* compound eye; arrangement of the corneal lenses (b), (c), and (d).

retinal photoreceptors and suggested that these were fiber optic light guides. This indicated that *Phacops* had developed an advanced optical mechanism to channel light to the retina for imaging in an oceanic environment, where the light levels were low and diffuse.

Levi-Setti (1975) was intrigued by trilobite eyes and their optics. He investigated the optics of their corneal lenses (Clarkson and Levi-Setti, 1975). They found that the *Phacops*'s lenses are highly biconvex and doubly refracting and that the shape of the lens duplicates the lens construction of Descartes (1637) in design, that is, an aplanatic lens making use of two Cartesian ovals. The lens was also almost an exact duplicate of a superb lens described by Huygens (1690), making use of a spherical first surface and a Cartesian second surface, hence intraocular lenses as depicted in Figure 12.27. Such a lens would correct for spherical aberration. Levi-Setti made a model of the trilobite lens and subjected it to different light conditions, finding that the lens would function as a light-gathering lens to optimize the light where there was little light in the environment.

It is not known whether the *Phacops* eye is a true compound eye or an aggregate of individual eyes since each eye functions separately to survey a different part of an object in space (like that of *Copilia*) or whether it is like more highly evolved imaging vertebrate eyes.

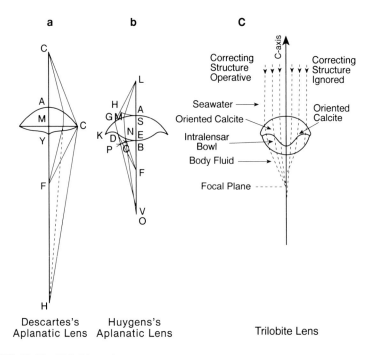

FIGURE 12.27 Trilobite, *Phacops,* lens optics compared to (a) Descartes's aplanatic lens and (b) Huygens's lens. (Courtesy of Prof. Levi-Setti, University of Chicago.) (c) Ray tracing through the lens of the trilobite eye. (From Levi-Setti, 1975.).

MOLLUSCS

In addition to insects and crustacea, molluscs, which include clams, oysters, nautiluses, octopi, and squids, are the most numerous among the invertebrates. Every kind of imaging eye is found among them from the pin-hole eye of *Nautilus* to the refracting-type eye of *Octopus* and *Squid* (Figure 12.28).

The cephalopod mollusc *Octopus vulgaris* has attracted considerable attention, for it is a large muscular animal that can grow to more than 10 feet in length. It lives in tropical and subtropical waters and is abundant in the Mediterranean. The octopus can achieve chameleon color matching of its surroundings, producing patterns by expanding and contracting the chromatophores in the skin.

The octopus eyes are unusually large compared to the size of the body. The brain behind the eye is one of the most highly developed among invertebrates. The brain is enclosed in a cartilaginous skull and is divided into fourteen lobes that govern different sets of functions; one lobe controls the jet apparatus, another lobe the memory, and so on, the optic lobes being the largest. All of its behavioral responses are primarily visual ones, therefore the octopus is an ideal animal for

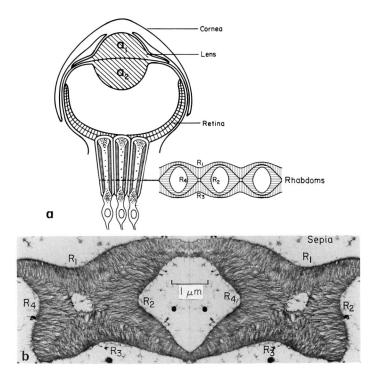

FIGURE 12.28   Mollusc cephalopod octopus eye. (a) The structure of the retina is formed of rhabdoms. Each rhabdom has four rhabdomeres (R_1–R_4). (b) Electron micrograph (from Wolken, 1975, 1986). The lens is constructed of two halves (a_1 and a_2).

studies of visual acuity; the eye–brain relationship has been extensively studied by J. Z. Young (1964) and Wells (1978).

The octopus uses both monocular and binocular vision. In monocular vision, the eyes face in opposite directions with their long axes roughly parallel. In binocular vision, the eyes are switched slightly forward relative to the body. Therefore, the octopus has a complete 360° field of view. The octopus can learn to distinguish one shape from another, even when the only difference between the two is their orientation. When housed in a laboratory tank, the octopus' freedom is not inhibited, for it can crawl out of its tank and roam around the room. Therefore, their eyes' optics are adapted to see in both water and air.

The octopus eye structure (Figure 12.28a) has been likened to a vertebrate eye with a transparent cornea, a lens, an iris, a diaphragm, and a retina. The large, spherical lens is ideal for maximizing light-gathering power in an aquatic environment. However, it differs fundamentally from the vertebrate eye in that the lens is formed by two joined hemispheres. The retinal photoreceptors are not rods and cones as in the vertebrate retina but rhabdomeres that form rhabdoms like those

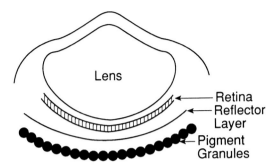

FIGURE 12.29   Scallop, *Pecten*, eye, schematic drawing; lens retina reflecting light layer and pigment granules.

described for arthropod compound eyes (Grenacher, 1879, 1886; Patten, 1887). Studies by Moody (1964) and Wolken (1971) of the retina of *Octopus vulgaris* and the closely related cuttlefish *Sepia officinalis,* as well as a study of the squid (Zonana, 1961), bear this out and show that the four rhabdomeres are radially arranged (Figure 12.28). The retina of *Octopus vulgaris* is at least 1 cm^2 in area with about 2×10^6 rhabdomeres per eye. The number of photoreceptors is roughly equivalent to the number of retinal rods of the vertebrate retina. The retinal photoreceptors are not inverted as in the vertebrate retina, so light reaches the photoreceptors directly.

The cephalopod *Sepia* is closely related to the *Octopus,* and its eye structure is similar to that of the octopus. In a cross-section of the rhabdom, four sides of the rhabdomeres are isolated by screening pigment cell granules. These pigment granules migrate depending on the light intensity in order to adapt to the light. The rhabdomeres measure about 70 μm in length and from 1 to 1.5 μm in diameter. In all the cross-sections, the lamellar structure is observed, whereas in all oblique and longitudinal sections the microtubular structure is seen. Here, as in the arthropod rhabdomeres, the microvilli's microtubules are about 500 Å in diameter and are separated by a double membrane about 100 Å in thickness.

In the scallop *Pecten maximus,* there are about sixty eyes distributed around the edges of the mantle, and each eye is about a millimeter in diameter. Scallops are phototactic and migrate, depending on the light intensity, either toward the light or toward darkness in search for an optimal environment. Their eyes detect movement of a shadow (or dark edges) across the animals' field of view. Images are resolved by reflection, due to the fact that scallop eyes (Figure 12.29) have shiny, hemispherical mirrors (Land, 1980).

SPECTRAL SENSITIVITY AND COLOR VISION

The question arises: Do arthropods (insects and crustacea) and molluscs "see" colors? For them to see colors, they need to have two different spectrally absorbing visual pigments, as cones do in the vertebrate eye. From numerous observations, we know that insects do see colors. To establish whether arthropods and molluscs do in fact see colors, research has been done on behavioral action spectra using ERG measurements which show the spectral sensitivity of the eye.

To precisely identify visual pigments requires chemical extraction and further purification (this is a tedious process requiring hundreds to thousands of eyes). This has now been circumvented by using a microspectrophotometer to directly obtain their photoreceptors' (rhabdom and rhabdomeres) absorption spectra. All such experimental data have shown that their visual pigments are rhodopsins that are remarkably chemically similar to rhodopsins in vertebrate eyes.

This brings us back to the question of how insects and other arthropods see colors. Insects exhibit considerable diversity in their photobehavior; many are diurnal and their eyes well-adapted for navigating in bright and dim light, while other insects are nocturnal and color blind (Wigglesworth, 1964).

Behavioral studies have shown that insect eyes can go from dim to bright light like vertebrate eyes with rods and cones. This implies that they should have different spectrally absorbing pigments.

Drosophila, in their flight patterns, go from high to low light levels with a Purkinje shift from photopic to scotopic vision, similar to the vertebrate eye going from rod to cone spectral sensitivity (Fingerman, 1952; Fingerman and Brown, 1952, 1953). This shift in sensitivity suggests that there are two spectrally different absorbing pigments in the photoreceptors of the *Drosophila* eye: one around 510 nm and the other in the ultraviolet. The photosensitivity response curves obtained by Fingerman and Brown (1952) indicated that the basic photosensitivity curve for *Drosophila* is that of the white-eyed mutant and that the response curves of other eye-color mutants differ from this only because of the screening effects of their eye-color pigments. The housefly, *Musca domestica,* like *Drosophila,* has a spectral sensitivity around 440 and 510 nm and in the ultraviolet around 340 to 390 nm.

Karl von Frisch (1914, 1950) pioneered the study of color vision in honeybees, *Apis mellifera,* and demonstrated that the bee could be trained to distinguish red, yellow, and green from blue and violet. Kühn (1927) also showed that the bee could distinguish the blue-green and near ultraviolet. These behavioral studies were confirmed and extended by Bertholf (1931), Hertz (1939), Daumer (1956), and Kuwabara (1957). They concluded that the primary ranges of bee sensitivity are in the near ultraviolet 300 to 400 nm, in the blue 400 to 500 nm, and in the yellow 500 to 600 nm. Goldsmith (1958, 1960), using electrophysiological methods, showed that there were several different absorbing visual pigments: one near 440 nm, another near 535 nm, and an ultraviolet sensitivity near 345 nm. It was then found that there were in fact three different visual pigments for the worker bee with spectral sensitivity peaks at 340, 430, and 530 nm (Figure 12.30) (Autrum and von Zwehl, 1962, 1964). Therefore, there are photoreceptors with maximum sensitivity for a green receptor, 543 nm, for a blue receptor, 440 nm, and for an ultraviolet receptor, 343 nm, in the worker bee (Menzel et al., 1986) which permit the bee to see colors.

The nocturnal moth, *Manduca sexta,* is also capable of seeing colors, for

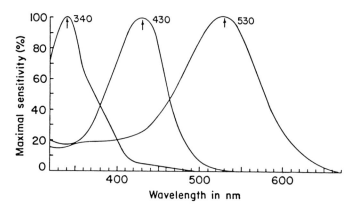

FIGURE 12.30   The worker bee absorption spectra.

microspectrophotometry has revealed three differently absorbing pigments with maxima around 450 nm, 530 nm, and in the near ultraviolet around 350 nm (Carlson, 1972). In other species of moths, *Deilephila elepenor,* maxima were found at 350 nm and from 440 to 460 nm. Other pigments that were isolated from their absorption spectra were around 520 to 530 nm, 460 to 480 nm, and 330 to 340 nm (Langer et al., 1986). These wavelength absorption peaks indicate that moths have a blue receptor, a green receptor, and an ultraviolet receptor. These wavelength absorption regions are similar to those found for the worker bee and imply that moths, like bees, can distinguish colors.

To clarify the dual spectral sensitivity and spectrally different absorption peaks, Autrum and Burkhardt (1961) and Burkhardt (1962) measured spectral sensitivity for the blowfly, *Calliphora erythrocephala,* using microelectrodes placed in single retinula cells. Three different spectral sensitivities were found in the visible with maxima at 470, 490, and 520 nm, and in the ultraviolet around 350 nm. In similar experiments with drone bees, Autrum and von Zwehl (1962, 1964) found two different receptors with maxima at about 340 and 447 nm. The 447 nm peak compares well with the observed spectral sensitivity maximum for drones at 430 nm and with the extracted photosensitive honeybee pigment, maximum near 440 nm (Goldsmith, 1958).

The blowfly eye-color mutant "chalky" lacks all eye-screening pigments. The rhabdom of the blowfly consists of eight rhabdomeres. Langer and Thorell (1966) found that for six of these rhabdomeres the absorption maximum was about 510 and for the seventh (asymmetric) rhabdomere about 470 nm. These spectral peaks come close to Burkhardt's maxima for spectral sensitivity, and presumably these absorption peaks are associated with two differently absorbing visual photopigments.

As for arachnids, spiders are of special interest in comparison to other insects, for it was thought for a long time that spiders can distinguish colors (Peckman and Peckman, 1887). In the wolf spider, only one visual pigment for the median eye could be identified, which absorbed maximally around 505 to 510 nm, but the greatest sensitivity was in the ultraviolet around 380, as with most insects. A more thorough study of the jumping spider, *Evarcha falca,* was determined from their behavioral action spectra, and Kästner (1950) demonstrated that spiders do have color vision.

Among crustacea, the lobster visual pigment has absorption peaks around 480 and 515 nm and the crayfish near 510 and 562 nm as compared to the honeybee and other insects near 430 and 530 nm; these absorption peaks indicate that they should see colors.

Screening pigments

The various eye colors seen in insects and crustacea come from different-colored pigment granules that surround the ommatidia. These pigment granules regulate the light that reaches the rhabdom. It was thought that these pigment granules may be chemically similar to the oil globule pigments found in the retinas of birds, turtles, lizards, and snakes, but there is no evidence to indicate that they are chemically identical.

The screening pigment granules have for the most part been identified as ommochromes, pterines, and pteridines (Ziegler, 1964, 1965; Ziegler-Gunder, 1956; Grossbach, 1957). Ommachrome pigments are yellow to dark red and fall into two classes: ommatines and ommines (Linzen, 1959). The ommachromes of the xanthommatin type are photosensitive, can be oxidized and reduced, and are pH-sensitive. Yoshida et al. (1967) suggested that these ommochromes could function like the quinones and the cytochromes in the electron–energy transfer chain.

In the eye of the housefly, *Musca domestica,* yellow pigment granules are found at the top of the ommatidium surrounding the corneal lens and the crystalline cone, whereas the red pigment granules are found closer to the rhabdom, its photoreceptor. Microspectrophotometry of the yellow and red granules showed that the yellow pigment granules have a broad absorption band maximum around 440 nm; the red pigment granules have a maximum absorption around 530 nm (Figure 12.31a) and a smaller peak near 390 nm (Strother and Casella, 1972). The yellow pigment absorption spectrum corresponds to the plant pigment xanthophyll, found in the fovea of the mammalian eye retina. The red pigment is pH-sensitive and shifts in its absorption peak from around 490 nm (in an alkaline solution) to 440 nm in acidic solution. This shift in absorption spectra is similar to that of rhodommatin, a red ommochrome-type pigment (a xanthommatin) isolated from insect eyes (Bowness and Wolken, 1959; Butenandt et al., 1954).

The blowfly, *Calliphora erythrocephala,* possesses eyes that have both yellow and red pigment granules (Figure 12.31b). The yellow granules absorb around 445 nm, while the red pigment granules have two absorption peaks around 380 nm and 540 nm (Langer, 1967). The yellow pigment absorption spectrum resembles that of an oxidized xanthommatin-protein complex, and the red pigment absorption

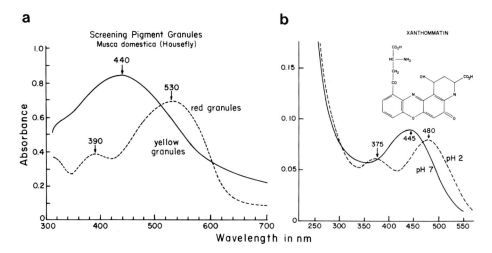

FIGURE 12.31 (a) The red and green pigment granules absorption spectra in the horsefly, *Musca,* eye compared to (b) absorption spectrum of xanthommatin.

spectrum is similar to the rhodommatin pigment (Butenandt et al., 1954; Butenandt and Neubert, 1955; Burkhardt, 1962). These spectra are also interesting from a physiological standpoint, for no obvious correlation between them and the overall spectral sensitivities for these insects is evident. Accordingly, Langer and Thorell (1966) made direct microspectrophotometric measurements of the *Calliphora* rhabdomeres within a rhabdom. They found two different spectra: one had two absorption peaks with maxima at about 380 and 510 nm and closely resembled the red screening pigment spectrum, and the other showed only a single peak near 470 nm and closely resembled the yellow pigment spectrum (Strother and Casella, 1972). In fact, photosensitive yellow pigments with an absorption peak near 440 nm were extracted from the housefly (Bowness and Wolken, 1959) and from the honeybee (Goldsmith, 1958), which closely match the yellow screening pigment absorption peak. Data obtained by Langer (1967) and Strother and Casella (1972) indicate that for the blowfly and housefly the combined effect of both yellow and red screening pigments is to effectively screen the separate ommatidia from light leakage of wavelengths from 320 nm to about 640 nm. It is thought possible that the yellow and red screening pigments are acting separately to screen two different visual pigments, namely, the one absorbing at 440 nm and the other near 510 nm.

The absorption spectra of these variously colored pigment granules suggest that those regions where the pigments permit passage of light coincide with regions of the spectrum where insects' visual pigments are most photosensitive. Thus, ultraviolet light from 350 to 400 nm and red light beyond 600 nm are transmitted and not absorbed by these pigment granules. This transmitted light is then available to the photoreceptor visual pigments and results in a greater sensitivity for the insect.

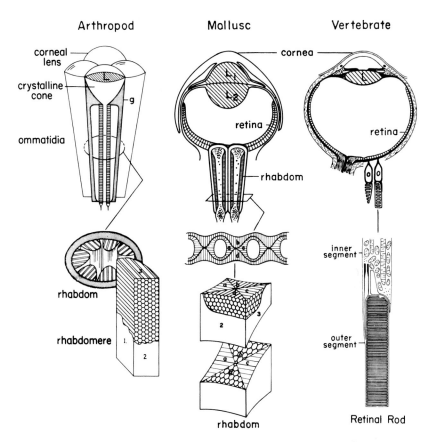

FIGURE 12.32   Phylogenetic structural relationships in eyes from arthropods to molluscs to vertebrate eyes.

CONCLUDING REMARKS

In reviewing the evolutionary development and structure of invertebrate eyes, no clear patterns could be discerned; most likely, different species evolved their eyes independently. For among invertebrate eyes, every known optical system device for forming an image has evolved, that is, from pinhole eyes to simple eyes to camera-type eyes to compound eyes and to eyes with refracting optics (Figure 12.32). In doing so, nature anticipated the development of modern optics.

CHAPTER THIRTEEN

Polarized Light in Nature: Detection by Animals

POLARIZED LIGHT

There are several optical phenomena in the sky that animals can use as compass cues. One is the direct light from the sun (a point source). In addition, the scattered light from the sky provides various kinds of information: light intensity gradients, spectral gradients, and a pattern of linearly polarized light. The main sources of linearly polarized light are sunlight, scattered by air molecules within the Earth's atmosphere, and light reflected from surfaces of water. Underwater light is also polarized, due to scattering by water molecules and particulate matter (Waterman, 1975).

Specifically, polarized light may be produced by Rayleigh scattering (atmospheric polarization), reflection (light scattering at the Brewster angle), dichroism (selective absorption in one direction), and birefringence (differing indices of refraction for the optical axes).

If we consider the behavior of the electric vector, known as the E-vector, as it interacts with matter, then the phenomenon of polarization may be analyzed. When light is produced, the E-vector of an individual wave can vibrate randomly in any direction. If the E-vectors of these waves vibrate in a single plane, then the light is said to be *plane polarized*. If the E-vector traces out a 360° path as it moves, the light is said to be *circularly polarized*. Each of these forms of polarization occurs in nature as the result of the interaction of the E-vector of light with matter (Können, 1985).

When light strikes an interface at an angle known as the Brewster angle, θ_B, the reflected light is polarized in the horizontal plane. This is the glare we notice from smooth surfaces when sunlight strikes these surfaces at certain angles. The Brew-

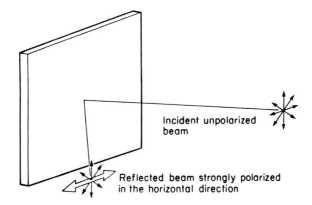

Incident unpolarized beam

Reflected beam strongly polarized in the horizontal direction

FIGURE 13.1   The production of a polarized beam of light by reflection. Note the large vector indicating a strong component of polarization in the horizontal direction. Not all of the incident light is polarized.

ster angle may be determined from the refractive indices of the surfaces as follows: $\tan\theta_B = n_i/n_t$. A schematic of the interaction and the direction of polarization is shown in Figure 13.1.

To reduce the glare caused by reflections from surfaces and to dim the brightness of a sunny, summer day, we employ sunglasses. These glasses are made of dichroic materials. Such materials have their molecules aligned in parallel rows and selectively absorb in one direction.

If polarized light is incident on dichroic materials (polarizers), then the transmitted intensity varies according to the law of Malus. This is given by, $I = I_o \cos 2\theta$, where θ is the angle between the direction of polarization of the light and the preferred direction of the polarizer. This phenomenon was accidentally discovered by the French physicist Etienne Malus in 1809; while looking at the sky through a crystal, Malus recognized that the light was polarized.

If a second polarizer (an analyzer) is now placed in the path of the light, the angle θ is now the angle between the preferred direction of the two polarizers (Figure 13.2). This permits the determination of the angle of polarization of a plane wave. If plane polarized light is incident on a birefringent crystal, the emerging light will be plane, elliptically, or circularly polarized, depending upon certain conditions. A crystal that is birefringent has two differing indices of refraction. If plane polarized light is incident on a crystal cut parallel to the optic axis, then the incident light splits into two rays called the *ordinary ray* and the *extraordinary ray*. If the optical path is defined as the index of refraction times the thickness (or distance of penetration), then the optical path of the ordinary ray is given as $n_o t$ and that for the extraordinary ray as $n_e t$. The path difference is $\Delta = (n_o - n_e)t$, and the phase difference is $\Delta = 2\pi/\lambda \, (n_o - n_e)t$ for these paths. For phase differences of 0, 2π, 4π, etc., the vibrations will emerge unchanged. For odd

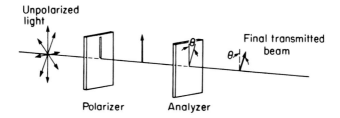

FIGURE 13.2   Relationship of the polarizer and analyzer for the detection of the degree of polarization of a beam of polarized light. The unpolarized beam is polarized by the polarizer, then sent through the analyzer. If the analyzer is rotated such that it is 90° to the polarizer, the beam of light will be extinguished and the direction of polarization may be determined from the measurement.

multiples of π, the vibration will be plane polarized at an angle of $2\ \theta$. For intermediate conditions, the light will emerge as elliptically polarized.

Two conditions must hold to create circularly polarized light: the amplitude of the ordinary and extraordinary rays must be equal, and the vector components of the incident E-vector must be equal ($\sin \theta = \cos \theta = 45°$). A commercial device that produces or detects circularly polarized light is called a *quarter wave plate* ($\lambda/4$), or *retarder,* since it introduces a delay or retards the light. Thin mica sheets cut into specific wavelengths are the usual material for commercial retarders. If plane polarized light is sent through a quarter wave plate, the emerging light is circularly polarized. If circularly polarized light is sent through a quarter wave plate, the emerging light is plane polarized. Comprehensive treatments of polarized light are given in Jenkins and White (1976) and Shurcliff (1962).

DETECTION OF POLARIZED LIGHT BY THE EYE

We have already stated that polarized light is present in the environment due to Rayleigh scattering and that the direction 90° from the Sun's position toward the horizon is strongly polarized. Since the direction of polarization indicates the relative position of the Sun with respect to the horizon, polarized light may be of use by animals as a directional compass for orientation and navigation.

Sir John Lubbock (1882), an English banker and naturalist, observed the behavior of ants and discovered that ants could detect ultraviolet light. He wondered how ants found their way to and from their nests while foraging for food. His observations led him to conclude that ants used the Sun as a compass for orientation and navigation. This seemed at this time to be a reasonable explanation, though it was soon questioned. For how could insects find their way when the Sun was hidden from view by clouds?

The question was not really answered until von Frisch's (1949) behavioral studies of the honeybee suggested that, in addition to the direction of a point

a   b   c

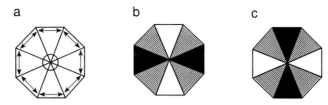

FIGURE 13.3   Rhabdomeres of rhabdom acting as an analyzer for polarized light. (a) The double arrows indicate the plane of vibration of the transmitted polarized light. (b,c) The Rhabdom responds differentially to polarized light. Opposite rhabdomeres are stimulated to the same extent but adjacent ones are stimulated differently. (After von Frisch, 1950, p. 107.)

source, the Sun, insects could utilize polarization information from a patch of blue sky. Even when the Sun is obscured by clouds, polarization cues are available. Thus, insects can make use of polarized light on cloudy days. The detection of polarized light requires an analyzer. The question arose: Where specifically is the analyzer for polarized light in the insect? Von Frisch (1950) and Autrum and Stumpf (1950) inferred from behavioral studies of honeybees that the polarized light detector was the eye and that the analyzer was built into the rhabdom structure. To test this hypothesis, von Frisch (1949, 1950, 1967) constructed a model based on the bee's rhabdom, using eight triangular polarizing elements, each transmitting a quantity of light proportional to the degree of polarization. In his model (Figure 13.3), opposite pairs of rhabdomeres would have their polarizers in parallel orientation as illustrated in our model (Figure 13.4).

In the eyes of insects, crustacea, and cephalopod molluses, the rhabdom is structured of rhabdomeres. Electron microscopy of the rhabdom shows a striking geometric arrangement of perpendicular and parallel microtubules (microvilli) of rhabdomeres that form the rhabdom and confirms von Frisch's structural model for detection of plane polarized light (Figure 13.3).

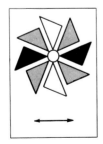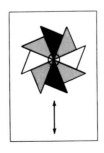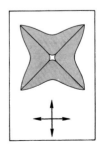

FIGURE 13.4   Model of rhabdoms in which their rhabdomeres form an open-type rhabdom and a closed-typed rhabdom and act as analyzers of polarized light. (From Wolken, 1986, refer to von Frisch model, Figure 13.3.)

The analysis of the detection of polarized light by these invertebrate is based on the dichroism of the photosensitive pigment within the rhabdomeres of the rhabdom. Dichroism, the selective absorption of the E-vector in one direction, occurs in the oriented visual pigment rhodopsin within the rhabdomere structure. Since molecules of rhodopsin are oriented within the rhabdomeres, they exhibit dichroic absorption and act as a polarization analyzer.

In the crayfish, *Orconectes virilis,* rhabdoms illuminated transversely show that their photosensitive absorption exhibits a dichroic ratio of 2 *in situ.* The major absorption axis matches the axial direction of the closely parallel microtubules of the rhabdomere. Since these microtubules are regularly oriented transversely in about twenty-four layers, with the axes of the microtubules perpendicular to each other in alternate layers, transverse illumination of a properly oriented rhabdom displays alternate dichroic and isotropic bands. If, in fact, the rhabdom constitutes a dichroic analyzer, its properties would depend on the arrangement of the microtubules in the individual rhabdomeres and on the orientation of the rhodopsin molecules, with their major axes parallel to the tubule direction and, hence, perpendicular to the normally incident illumination. The explanation of the polarized light analyzer action in the rhabdom is that the absorbing dipoles of the rhodopsin molecules, as in the vertebrate retinal rods, lie parallel to the membrane surface but are otherwise randomly oriented (Waterman et al., 1969).

Burkhardt and Wendler (1960) recorded action potentials of single retinula cells in the compound eye of the blowfly, *Calliphora,* with intracellular electrodes. They found that rotating the plane of polarization resulted in a 50% difference in amplitude between maximum and minimum responses. This effect was obtained with blue and white light but was not observed when red light was used as a stimulus. Their finding, that blue and white light had effects on the receptor potential but that red light had none, may have its explanation in the dichroic absorption of oriented rhodopsin in the rhabdomere.

The above model for polarization sensitivity developed from the work of von Frisch and supports his behavioral studies and the microscopic and electrophysiological measurements. This model is based on dichroic absorption and requires an oriented analyzer. Such an analyzer has been shown to exist in oriented rhodopsin, as well as in oriented microtubules. There are, however, other models which utilize the experimental data, but they vary in their explanations of polarization detection from that of the von Frisch model.

Rossel and Wehner (1986) suggested an alternative explanation derived from their investigations of how honeybees detect polarized light, and there is some evidence to support their model. In our laboratory we found that ants (*Tapinoma sessile* and *Solenopsis saevissimmae*), the housefly (*Musca domestica*), the firefly (*Photurus pennsylvanicus*), and the Japanese beetle (*Popillia japonica*) orient to plane polarized light under circumstances which prevent the use of background reflections as cues for orientation. These insects had a compass reaction (the source of stimulation serves as a fixed point by which an orientated path is maintained) ±45° with respect to the plane of polarization of the incident light

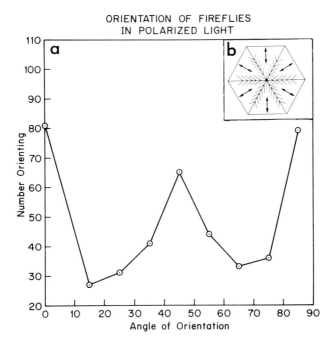

FIGURE 13.5   Orientation of fireflies to polarized light.

beam (Figure 13.5). All oriented at 0° and 90° except the Japanese beetle (Marak and Wolken, 1965).

The use of a white background which reflects normally incident polarized light in a circular pattern and a black background which reflects elliptically led to a consistent difference in the response curve. With the white background, there was more orientation at 45° and less at 0° and 90°. This response difference is consistent with the hypothesis that there are more cues for simple phototactic responses on the black background which reflects more light that is perpendicular rather than parallel to the plane of polarization. The reflectance pattern of nonpolarized light is circular, while the reflectance pattern of polarized light is elliptical, with the long axis perpendicular to the plane of polarization. In normal daylight, the long axis of the ellipse will point to the solar azimuth.

Another hypothesis, advanced by Baylor and Smith (1958), was that the direction of the E-vector of a beam of plane polarized light could be perceived through a simple intensity discrimination because the direction of vibration is resolved into intensity gradients by reflection from the background. In fact, reflection patterns from the environment do resolve polarized light into patterns of graded intensity. Kennedy and Baylor (1961) suggested that the compound eye merely discriminates intensity and is not a direct analyzer. In this context, an interesting observation by von Frisch (1967) showed that when the bee was dancing in the hive, the lower parts of the eye would be used in perceiving reflection patterns, while the upper parts of the eye would be used if the stimulus was the direct perception of polarized light from the sky. Von Frisch found that masking even a small area of the upper part of the eye disrupted dancing, while masking the lower part of the eye had no effect.

Houseflies and fireflies appear to orient to polarized light only within a narrow range of frequencies, suggesting that there are two photoreceptors with different absorption E-vectors. If these two receptors are oriented at 90° with respect to each other, and if the long axis of the visual pigment molecule is oriented in a single plane but free to rotate in that plane, the receptors will absorb polarized light in a ration of two to one when the E-vector is parallel to the absorption plane of one of the photoreceptors.

Stephens et al. (1953) advanced a model for polarization analysis which is dependent upon reflection and refraction at the air–corneal interface. This led to the examination of the structure of the corneal lens and crystalline cone as directional polarizers of light. Electron microscopy of the corneal lens of the firefly (*Photurus pennsylvanicus*) shows that it is formed of layers (lamellae) and that the lamellae form a single or double spiral in transverse sections. In each successive layer, the fibril direction is rotated on an axis perpendicular to the planes. The direction of the fibrils rotates through 180° from one lamella to the next, forming helices (Neville, 1975; Neville and Caveny, 1969). This was also observed in the corneal lens of the June beetle (*Scarab phyllophaga*), the carpenter ant (*Camponatus herculenus*), the fruitfly (*Drosophila melanogaster*), and the housefly (*Musca domestica*). A structural model for the corneal lenses of these insects is illustrated in Figure 12.13. The spacing between lamellae in the corneal lens is of the right order to function as a polarizer, whereas the structure of the rhabdomeres that form the rhabdom is the analyzer for the direction of the polarized light (Brown and Wolken, 1979). Thus a polarizer and analyzer are built to the eye.

A number of other interpretations can be found for the detection and use of polarized light and the different optical methods that evolved in animals. These may be related to the environmental niche each organism inhabits. It is also possible that some organisms make use of one or more of the above suggested methods for polarized light analysis. What is crucial is that these invertebrates are capable of using plane polarized light for orientation and navigation.

Plane polarized light is the dominant form of polarized light in the natural environment. Circularly polarized light also occurs to a limited extent, as is detected in scarab beetles (Konnen, 1985). A. A. Michelson (known for his studies of optics and with Morley for experimentally measuring the speed of light) became interested in the optics of the metallic color reflections from the cuticle of beetles. Michelson (1911, 1927) found that the light reflected from the beetle, *Plusiotis resplendens,* was circularly polarized as was the light from other beetles, butterfly wings, and feathers of birds. He concluded that "the effect must therefore be due to a screw structure of ultra-microscopic molecular dimensions."

Conmar Robinson (1966), working on the chemical structure and optical prop-

erties of liquid crystals, became fascinated by the studies of Michelson and obtained a variety of beetles from the British Museum of National History, with
which he repeated Michelson's polarization measurements. In doing so he used a
quarter-wavelength retarder to determine whether the reflected light from these
beetles was circularly polarized and observed that the reflected light was indeed
circularly polarized. Robinson stated: "It would be of interest to consider what
survival value can account for the occurrence of this most unusual property in so
many species." The experimental observations that circulatory polarized light is
reflected from the cuticle of scarab beetles or other insects raises the question as to
whether the beetle eye is able to detect the circulatory polarized light that is
reflected from the cuticle. We became interested in Robinson's observations when
we were studying the beetle eye corneal lenses optics and their photoreceptor
rhabdom structures so it was natural to inquire whether beetles can use the information from the reflected circulatory polarized light for orientation. In our laboratory, by using a simple retarder, wavelength 540 nm, and a polarizer, we found
that the light reflected from the June beetle (*Scarab phyllophaga*) and Japanese
beetle (*Popillia japonica*) was circularly polarized. Beetles are normally of bright
metallic colors but, under these conditions, appear black as the circularly polarized
light is extinguished by the polarizer. Whether the eyes of these beetles are able to
detect the circularly polarized light that is reflected from their cuticles was not
certain. If beetles do detect circulatory polarized light, they need to possess a
retarder that changes circularly polarized light to plane polarized light. The corneal
lenses of many beetles are birefringent (Meyer-Rochon, 1973), but it is not known
whether the corneal lens functions as a retarder. If such an optical system is
present, it would permit the beetle to detect circularly polarized light and use it for
orientation.

CONCLUDING REMARKS

The analysis of polarized light sensitivity in arthropods has been well-documented
since the early studies of von Frisch. In vertebrates, there is experimental evidence
that birds can detect the polarized light scattered in the atmosphere and can orient
to cues from skylight polarization (Able and Able, 1993).

Helbig and Wietschko (1989) called attention to avian orientational cues from
skylight polarization. At sunrise and sunset, a band of maximally polarized light
with its E-vector perpendicular to the Sun runs through the zenith at 90° degrees
from the Sun. This pattern remains detectable through a polarizer up to 45 min
after sunset and provides a cue for birds to perceive polarized light to its E-vector
direction.

Under water, the light field is polarized due to the scattering of water molecules
and particulate mater. Fish use their eyes to detect the polarized light (Waterman,
1984, 1989). Hawryshyn (1992) investigated the mechanism by which fish detect
polarized light. He found that certain species of fish will orient themselves in

response to a particular angle of polarization. In addition, if the polarized light included ultraviolet light (around 380 nm), the fish were able to orient relative to the polarized field accurately, but when ultraviolet light was absent in the stimulus, the fish did not orient to the E-vector. His finding that ultraviolet light is necessary for the accurate perception of the angle of polarized light indicates that fish possess an ultraviolet-sensitive visual pigment in the retinal photoreceptors of their eyes.

The question arises whether the human eye can detect polarized light from the environment. Haidinger (1844) discovered that polarized light can be detected by the human eye. According to Haidinger, if one gazes for a few seconds at a clear field of white light that is linearly polarized with the electric vibration horizontal, then glances at a clear field in which the direction is vertical, a faint pattern is seen consisting of a small yellowish brush with intervening areas bluish. This can sometimes be observed when we stare at a cloudless sky at an angle of 90° from the sun, that is linearly polarized and then glance at a clear field.

Shurcliff (1955, 1962) reinvestigated this phenomenon and found circularly polarized fields also produce brushes at different and distinctive azimuths, so that merely by eye an observer can tell the handedness of polarization. Right-circularly polarized light produces (for most observers) a brush. He presumed that linear dichroism occurs in the yellow pigmented macula of the eye, bringing about birefringence effects in the refracting cornea and lens.

These observations are interesting and need to be reinvestigated. We are still left with the unresolved question of whether the human eye can detect polarized light.

CHAPTER FOURTEEN

Light That Controls Behavior: Extraocular Photoreception

There are many creatures which have no eyes (as we understand the term) and yet "see" (using the word in its widest sense). . . .
—SIR STEWARD DUKE-ELDER, 1958, *The Eye in Evolution*

EXTRAOCULAR PHOTORECEPTION

Eyes are not the sole means of photoreception, for photosensors are located over the general body surface. As a result eyeless and blind animals can sense light. This diffuse photosensitivity over the whole or parts of the animal's skin is described as the *dermal light sense*. But, even deeper tissues in the body, such as neural and brain cells, are remarkably photosensitive. The photoreceptors that are outside the eye are referred to as *extraocular* or *extraretinal*. The research into extraocular photoreception for all types of photoreception that is not initiated through the eye was pioneered by Steven (1963) and Millott (1968, 1978).

Extraocular photoreception plays an important role in the behavior of animals. How such a photoreceptor system functions either alone or in conjunction with the visual system is of considerable interest. Many questions remain concerning the extraocular photoreceptor system. Where are these photoreceptor cells precisely located in the animal's body, what is their structure, and what is or are the photoreceptor pigment molecules? In finding answers to these questions, we will be better able to determine the effect of the extraocular photoreceptor system on animal behavior.

Behavioral responses

In animals, distinct behavioral responses are associated with extraocular photo-reception. These responses for the most part are observed as phototactic movements toward or away from a light source. The movement may be in bending, in the contraction of a part or parts of the body, or in the movement of the whole animal. Other types of behavior include migration, circadian rhythms, and sexual reproductive cycles.

Many invertebrates respond to a sudden change in light intensity by a withdrawal reaction. Steven (1963) attributed such photobehavior to a *dermal light sense*. Millott (1968) preferred the all-inclusive term "extraocular" for all types of photobehavior that were not initiated through the eye. The shadow response of the sea urchin *Diadema* is a withdrawal from the light, which is accompanied by a complex spine waving reaction. In annelids, the tail is withdrawn; in *Nereis diversicolor,* the photosensitive areas are located on the parapodia and pro- and peristromium (Gwilliam, 1969). In nematodes, there is a phototactic response to light but many lack localized photoreceptors (Croll et al., 1975). The marine worm, *Golfingia gouldii,* reacts to light by a retraction of the proboscis (O'Benar and Matsumoto, 1976). The burrowing sea anemone, *Calamactis praelongus,* bends toward the light (Marks, 1976). The adult sea squirt, *Ciona intestinalis,* orients in the direction of the light, accompanied by the opening and closing of its siphons (Dilly and Wolken, 1973). The *Ciona* body surface is also sensitive to changes in the light intensity, but not all parts of the body are equally sensitive. The most light-sensitive area is found in the region of the ganglion cells. Responses to changes in the light intensity are localized contractions or total contraction of the body. These contractions and elongations continue in *Ciona* even when the siphons are removed.

In many marine animals, extraocular photosensitivity is associated with migration. Migrating rhythmic behavior has been closely studied in the squid, *Todarodes sagittatus,* and in the tubellarian, *Convoluta roscoffensis. Convoluta* lives in the sandy bottoms during high tides and emerges onto the surface of the sand during low tides. *Convoluta* is truly an animal-plant; it lives in symbiosis with algae and searches for light to do photosynthesis. When *Convoluta* are brought into the laboratory and placed in constant light, vertical migration continues in synchronization with the timing of the tides. Such rhythmic behavior takes place only in constant light but not in constant darkness (Palmer, 1974; Wolken, 1975).

The extraocular system is also involved in the entrainment of circadian rhythms (Bennett, 1979; Hisano et al., 1972a). While such behavior can function in the absence of eyes, it often operates in conjunction with the visual system. For example, in *Aplysia,* the daily rhythmic behavior is not dependent on the eye, though the rhythmic activity can be modulated by the eye (Block and Lickey, 1973). The extraocular photoreceptor system directly mediates the entrainment of the circadian oscillator via the sixth abdominal ganglion.

This circadian oscillator modulates the frequency of the spontaneously active neuron labeled R15. The R15 neuron can be entrained to a light cycle after removal of the eye. However, the interaction between the R15 neuron and the eye is complex (Lickey et al., 1976; Lickey and Zack, 1973). The rhinophore in *Aplysia* is a peripheral nerve and unique in that it is sensitive to light as well as to chemical and tactile stimuli. A circadian oscillator exists in this nerve, that is, it was found to be involved in circadian activity (Chase, 1979). In the horseshoe crab, *Limulus,* there are a number of light-sensitive, extraocular receptor sites that mediate circadian behavior (Barlow, 1986, 1990).

In the garden slug, *Limax maximus,* the extraocular receptor system measures the duration of increasing daylight. The increasing length of light during the day results in the secretion of a "maturation hormone" by the brain and in turn initiates reproductive development (McCrone and Sokolove, 1979).

Truman (1976) investigated extraretinal photoreception and circadian rhythmic behavior in insects. In response to light, some insects exhibit phototactic flight movements toward light, either to the blue or the red region of the visible spectrum. For example, the alfalfa weevil, *Hypera postica,* orients toward the red region of the spectrum (Meyer, 1977). In the grasshopper and the silk moth, entrainment of flight rhythms occurs (Dumortier, 1972). Light also influences hormonal-reproductive behavior as observed in some species of insects; in the pupae, termination of diapause will occur when exposed to a photoperiodic regime.

Extraocular photoreceptors in vertebrates, amphibians, reptiles, fish, birds, and mammals measure light intensity and were found to function in orientation, in circadian rhythms, and in determining the timing of reproductive sexual cycles (Wurtman, 1975). For example, light-induced backward swimming (negative phototaxis) has been demonstrated for the eel, *Anguilla anguilla* (Van Veen et al., 1976). The orientation of salamanders and frogs to new compass directions can be imposed by an altered light–dark regime, but when their heads are covered, the orientation does not occur (Adler and Taylor, 1973; Taylor, 1972; Taylor and Ferguson, 1970). In eyeless larval bullfrogs, extraocular photoreceptors serve for perception of celestial cues and for spatial orientation. In lizards, entrainment of activity patterns such as circadian locomotor activity, gonadal responses, and color changes occurs (Underwood, 1977). The extraocular photoreceptors of the pineal organ and the frontal organ are capable of perceiving photoperiodic changes for synchronization of their "biological clocks" (Justis and Taylor, 1976).

In birds the extraocular photoreceptors function to control rhythmic flight behavior and the reproductive system as well (Hisano et al., 1972c; McMillan et al., 1975a,b; Menaker and Underwood, 1976). Newly hatched pigeon chicks, *Columba livia,* when they are hooded to prevent light from reaching their eyes, and then exposed to light will bring about head wagging and leg extension. The extraocular system also functions in rhythmic perching, but the control of the gonadal response can function alone or together with the visual system (McMillan et al., 1975a,b).

These few examples of extraocular behavioral responses are, for the most part, simple reflexes, but when they involve neural, brain, and hormonal systems, these mechanisms are very complex.

Where are and what are the photoreceptor structures?

The behavioral responses of extraocular photoreception are now recognized, but the precise identity and structure of these photoreceptors are more difficult to determine. The structure of the extraocular photoreceptors may be similar to, or very different from, the retinal photoreceptors of the eye. In many animals, the extraocular photosensitive areas are widespread, and not all areas are equally sensitive. Therefore, the dermal photosensitivity is diffuse and the receptors may or may not be confined to certain areas of the skin. For example, the tail of larval lampreys is more photosensitive than the surface of the skin; in sea squirts and in some bivalve molluscs, the photosensitivity is located in their siphons. In the gastropod mollusc, *Onchidium,* the photoreceptors are in the periphery of the mantle (Hisano et al., 1972a,b). In the sea urchin, *Diadema,* the skin is pervaded by nerve fibers that are most photosensitive (Millott, 1978). In annelids, such as the polychaete *Nereis diversicolor,* the photosensitive areas are located on the parapodia and pro- and peristromium. The dermal or diffuse photosensitivity for most vertebrates is associated with immature animals (embryos, chicks, and newly born rats). For the frog, the dermal response to light has been obtained from sections of the frog skin and the skin over the frontal organ (Eldred and Nolte, 1978). In pigeon chicks, the dermal light response is located in the skin as demonstrated for opaque, caped pigeons (Harth and Heaton, 1973).

In contrast to the dermal photosensitivity, a receptor site for extraocular photoreception is found in neural tissue among a wide range of animals. As an example, a receptor site is located in the cerebral lobes of the alfalfa weevil, *Hypera postica* (Meyer, 1977). Receptor sites are also found in various ganglia (sub- and supraoesophageal and sixth abdominal ganglia) of the crayfish, *Procambarus clarkii,* and the scoprion, *Heterometrus fulvipis* (Geethabali and Rao, 1973; Hisano et al., 1972a,b,c; Larimer et al., 1966). In the hardshell clam, *Mercenaria,* a structure is located in the fine distal processes of the axons. These axonal processes contain a photosensitive pigment; this suggests a receptor site because it is found in an organized, membranous, pentalamellar structure (Weiderhold et al., 1973). Within the sixth abdominal ganglion, the photosensitive region of the sea hare, *Aplysia,* there are yellow-orange pigmented granules, *lipochondria,* which can be isolated from the cytoplasm of these neural cells. The lipochondria granules are crystalline, membrane-bound structures and are light-sensitive. The lipochondria granules (when briefly illuminated for about 30 sec) undergo a structural change accompanied by the release of calcium (Kraughs et al., 1977; Brown et al., 1975). Also, in *Aplysia,* the peripheral nerve, or rhinophore, is an extraocular photoreceptor site that has been found to control both the locomotor and the circadian rhythmic activities (Chase, 1979).

In the cephalopod molluscs squid and octopus, the extraocular photoreceptors have been identified in the paraolfactory vesicles and stellate ganglia. The paraolfactory vesicles are found near the optic tract and are connected to the brain in parallel with the eyes. Their photoreceptors are structured of microvilli, similar to the visual photoreceptor rhabdomeres of arthropod eyes (Baumann et al., 1970; Mauro and Sten-Knudsen, 1972; Wolken, 1971, 1975, 1986).

The passage of light through the skull was not considered to be related to extraocular photoreception until it was demonstrated that light does penetrate into the brain (Van Brunt et al., 1964). The pineal organ and parapineal region lie near the surface of the brain and are light-sensitive. It was found that the parapineal parietal eye in lizards, the stirnorgan in the frog, and the pineal in birds, sharks, and mammals are the extraocular photoreceptors that influence circadian rhythms. The Hardian gland is closely connected with deeper brain structures. In the rat, it is associated with the circadian rhythm (Wetterberg et al., 1970a,b,c). Other regions of the brain, such as the hypothalamus, pituitary, and rhinencephalon, are also sites for extraocular photoreception that mediates gonadal responses (Hisano et al., 1972c; Van Veen et al., 1976).

Electrophysiology

The extraocular, dermal, and neural cellular areas of animals can be explored by electrophysiological methods, using microelectrodes to probe these cells and to determine whether measurable electrical signals can be recorded in response to light.

Electrophysiological data have been obtained from the neural cells of invertebrates. In these animals, photosensitive neurons exhibit spontaneous electrical activity and transmit excitatory and inhibitory signals in response to light. The spontaneously active cells that are inhibitory respond to light by either partial or complete cessation of ongoing electrical activity. Neurons exhibiting such responses occur in the sixth abdominal ganglion of the sea hare, *Aplysia,* in the snail, *Onchidium verraculatum,* the scorpions *Heterometrus fulvipis* and *Heterometrus gravimanus,* and in the cerebral ganglion of the marine worm, *Golfingia gouldii* (Andersen and Brown, 1979; Hisano et al., 1972b; Geethabali and Rao, 1973; O'Benar and Matsumoto, 1976). For the spontaneously active cells that are excitatory, spike frequency increases in response to light. Photosensitive neurons in the snail, *Onchidium verraculatum,* show both excitatory and inhibitory responses. Neurons that are not spontaneously active become active only upon illumination (Hisano et al., 1972b). In *Aplysia* giant neurons, excitatory responses were found to light at 579 nm (Chalazonites, 1964). Excitatory responses to light were also recorded from cells in the paraolfactory vesicles of the squid, *Todarodes sagittatus,* and from cells in the stellate ganglion of the octopus, *Eledone moschata* (Mauro and Baumann, 1968; Mauro and Sten-Knudsen, 1972). Spontaneously active cells may also show *on* and *off* discharges in response to onset and cessation of illumination. The spontaneously active photoresponsive cells in the cerebral ganglion of the marine worm, *Golfingia gouldii*, exhibit this response (O'Benar and Matsumoto, 1976).

The latency of response to illumination may be prolonged as in the crayfish, *Procambarus*. For the crayfish, an excitatory response follows a prolonged latent period, and the excitatory response is followed, in turn, by marked discharges (Bruno and Kennedy, 1962). A decrease in the latency of response also occurs as the intensity of illumination increases. This decrease in latency is a way of coding for the change in intensity. Such latency changes have been reported for the photosensitive cells of the scorpions *Heterometrus fulvipis* and *Heterometrus gravimanus*, the marine worm, *Golfingia gouldii*, and the squid, *Todarodes sagittatus* (Geethabali and Rao, 1973; Mauro and Sten-Knudsen, 1972; O'Benar and Matsumoto, 1976). In addition to neural discharges, generator potentials also occur in photosensitive neurons. For example, generator potentials were recorded from cells in the paraolfactory vesicles of *Todarodes sagittatus*, the stellate ganglion of *Eledone moschata* and the caudal photoreceptor of *Procambarus* (Mauro and Baumann, 1968; Mauro and Sten-Knudsen, 1972; Wilkens and Larimer 1972).

Only in a few animals has the threshold for light excitation been measured, but the experimental results are of interest. The pallial nerve of the surf clam, *Mercenaria*, and the nerve from the pineal have light intensity thresholds of the order 6 $\times$ 10^{-9} $\mu W/cm^2$. The abdominal ganglia of the sea hare, *Aplysia*, and the crayfish, *Procambarus*, and the stellate ganglion cells of the octopus have light intensity thresholds of the order of 4 $\times$ 10^{-4} $\mu W/cm^2$ (Anderson and Brown, 1979). These values for threshold excitation are comparable to the excitation for insect eyes but less sensitive to that required for the vertebrate eye (Mauro and Bauman, 1968).

For vertebrates, the regions of extraocular photosensitivity are the pineal and extrapineal areas (parietal eye or frontal organ). The electrophysiology of the pineal region has been studied in a variety of animals, e.g., the shark, the lizard, and the frog. In the spotted dogfish shark, *Scyliorhinus caniculus* L., the pineal is sensitive to light, giving a positive slow wave accompanied by inhibition of spontaneous activity upon illumination. With continuous illumination spike activity is strongly depressed, followed by adaptation and some spike activity returns. When light intensity is increased, the latency for onset of inhibition also increases. Removing the light stimulus results in recovery (Hamasaki and Streck, 1971).

The parietal eye in the lizard is part of the pineal complex and is connected to the pineal by the parietal nerve (Figure 14.2). The parietal eye cells are spontaneously active during daylight and send afferent impulses to the pineal organ. Within the pineal, the efferent neurons to the parietal eye have different sensitivities to light and darkness. This sensitivity is mediated by two different neurotransmitter molecules. In the light, the efferent neurons are sensitive to norepinephrine and conduct impulses back to the parietal eye, enhancing its response to light. In the dark or at night, the parietal eye generates impulses to the pineal efferents which are now most sensitive to serotonin (Engbretson and Lent, 1976).

In the frog brain, the diencephalon exhibits spontaneous electrical activity in darkness but inhibits this spontaneous electrical activity in the light, indicating that the diencephalon is photosensitive (Dodt and Jacobson, 1963). In the frontal organ (part of the pineal complex), investigators have found that ultraviolet light at 355 nm will cause a negative slow potential with inhibition of spike activity, but green light at 515 nm will cause a positive slow potential with increased spike activity (Eldred and Nolte, 1978). These electrophysiological data would indicate a photosystem that involves a pigment with two active physiological states. Rayport and Wald (1978) measured the electrical responses and determined the electrodermogram (EDG) of isolated pieces of frog skin taken from the skin over the frontal organ. They found that the EDG showed hyperpolarization at low skin resting potentials and depolarization at high resting potentials. The maximum wavelength for excitation was found to be around 385 nm, and repeated exposures led to adaptation. However, exposure to longer wavelengths of about 500 nm led to recovery. They concluded that a photoreversible pigment system exists in the frog skin frontal organ involving two active states. These experimental data taken together suggest that the photoreceptor pidment molecule, like that in the retina of the eye, is a rhodopsin system.

Spontaneous activity appears to be a feature of cells involved in vertebrate as well as invertebrate extraretinal photoreception. Inhibition of activity occurs when light is present or directed at spontaneously active cells. In some species, as in the lizard, photoresponsiveness is enhanced via a feedback system, while in others adaptation occurs, and the latency changes with increased light intensity, as observed for the dogfish shark.

The electrophysiological data for the animals studied indicate that neural cells are capable of detecting and transducing light stimuli independent of an existing visual system. Such cells exhibit the classic responses of photosensitive cells, such as hyperpolarization, generator potentials, and spike discharges. Both excitation and inhibition are evident, as well as responses that code for light intensity changes, onset and cessation of stimulus, circadian activity, and adaptation and recovery. In the integrated animal, where mutual interactions with the visual and other sensory systems are occurring, the electrophysiological data are more difficult to interpret.

Extraocular spectral sensitivity and the photoreceptor pigment

Extraocular spectral sensitivity responses have been primarily determined from behavioral action spectra and electrophysiological measurements. The wavelengths of light that produce the response have been found to be in the blue (from about 400 to 500 nm), and other photoresponses have been found to occur in the near-ultraviolet as well as in the infrared regions of the spectrum.

For most insects, the behavioral action spectrum sensitivity peak is around 450 nm (Truman, 1976). Spectral sensitivity in the red region of the spectrum has been noted in the alfalfa weevil, *Hypera postica,* which responds from 650 nm and further into the red, and in the butterfly, which responds near 610 nm (Meyer,

1977). In two species of scorpions, *Heterometrus fulvipis* and *Heterometrus gravimanus,* where the pigmented tail segment and telsonic nerves are photosensitive, the spectral sensitivity was found to have two peaks—one around 586 nm and the other near 440 nm—but the pigment was not identified (Geethabali and Rao, 1973). In the silkworm, *Bombyx mori,* the photoperiodic receptors of the brain were identified as retinal and 3-hydroxyretinal, chromophores of insect visual pigments (Hasegowa and Shimizu, 1988).

In the sea hare, *Aplysia,* the light-sensitive neurons have pigment granules, or lipochondria, whose major absorption peak is around 490 nm, indicating that these pigment granules contain a carotenoid. The pigment granules also contain another pigment with an absorption peak around 579 nm (in the oxidized state) and a heme-protein identified as the photoreceptor molecule (Austin et al., 1967). Other researchers have found a carotene-protein with absorption peaks at 463 and 490 nm and a heme-protein with absorption peaks at 418, 529, and 542 nm (Chalazonitis, 1964; Gotow, 1975; Hisano et al., 1972c).

The gastropod mollusc, *Onchidium verraculatum,* possesses orange-pigmented neurons in the sub- and supra-oesophageal ganglia that are light-sensitive. Extraction of the neural tissue isolated a red pigment which was identified as a heme-protein and a yellow pigment which was identified as a carotenoid. Similar evidence for a heme-protein and a carotene-protein was obtained from the orange-pigmented neural tissue of the snail, *Lymnaen stagnalis* (Benjamin and Walker, 1972). In cephalopod molluscs, as previously mentioned, the extraocular photoreceptors are the well-developed paraolfactory vesicles that are closely associated with the brain. Rhodopsins were isolated from the paraolfactory vesicles of the squid and from the epistellar body on the surface of the stellate ganglia of the octopus (Baumann et al., 1970; Mauro, 1977; Mauro and Baumann, 1968). In the deep-sea squid, *Todarodes pacificus,* both a retinochrome and rhodopsin have been isolated and identified from the paraolfactory vesicles. The retinochrome is associated with myelin-like lamellar structures in the cytoplasm, and rhodopsin is found in the photoreceptor microvilli membranes of the parolfactory vesicles (Hara and Hara, 1980).

In the hard-shell clam, *Mercenaria mercenaria,* the spectral sensitivity peak in excitation of the nerve bundles is around 510 nm, suggesting that the photoreceptor pigment is a rhodopsin (Weiderhold et al., 1973). In nematodes, *Chromadorina viridis,* the spectral response sensitivity was found near the ultraviolet around 366 nm. The photoreceptor in the response is believed to be located in the oesophageal musculature. The pigment found in the structure was a heme or a heme derivative, and a rhodopsin was not identified as their photoreceptor pigment molecule (Croll, 1966).

For vertebrates, extraocular photoreceptors are found in the skin, in nervous tissue cells, and in the pineal of the brain. The identification of the photoreceptor pigment molecules has come from spectral sensitivity and electrophysiological measurements. For example, the spectral data of the frog skin was at 385 and 500 nm and of the frog pineal organ 355 and 515 nm. These spectra are strikingly similar and most likely the pigment is a rhodopsin molecule (Rayport and Wald,

1978). Additional evidence for a rhodopsin photosystem was obtained from spectral sensitivity measurements of around 506 nm for the dark-adapted frog pineal and of around 500 nm for the pineal of the small spotted dogfish shark, *Scyliorhinus caniculus* (Eldred and Nolte, 1978; Hartwig and Baumann, 1974). The Hardian gland in the brain is also associated with extraocular photoreception. A reddish-brown pigment was extracted from mouse and rat Hardian glands that was identified as a porphyrin (Wetterberg et al., 1970a, Watanabe, 1980). Nevertheless, it has been indicated that the photoreceptor pigment molecule for extraocular photoreception—obtained from behavioral action spectra, spectral sensitivity, and the spectra of pigment extracted from photoreceptor areas—is a rhodopsin, the visual pigment of animal eyes. Other pigments—including carotenoids, porphyrins (hemes and cytochromes), and flavins—have been implicated. Most likely, these pigments function either alone or together with rhodopsin in the photoprocesses of extraocular photoreception.

The pineal: an extraocular photoreceptor

> *Man is, among other things, a remarkably living sun dial.*
> —LYALL WATSON, 1973, *SuperNature*

The pineal organ (gland) in the brain is associated with extraocular photoreception. The pineal has fascinated humans for a very long time, its functional role in the brain being a mystery. The pineal was thought of as an "eye" by the Hindus in India, as expressed in their literature of enlightenment. René Descartes (1637) believed that the pineal was the "seat of the soul" and visualized that the events of our world are perceived through the eyes by a series of fibers to the pineal in the brain (Figure 14.1). The pineal was also thought of as our "third eye." Questions remained, however, concerning evolution and the role of the pineal gland in the brains of animals. Early evolutionists thought that the pineal was a vestigial organ of our reptilian ancestry, and as a result, the investigation of its function was neglected for some time. Zrenner (1985) has reviewed these interesting early historical accounts of the pineal. We can now turn to more recent investigations of the pineal organ and its structure, chemistry, and function in animal behavior.

The pineal organ is a mass of cells located at the base of the brain near the top of the spinal column (Figure 14.3a,b). It is a small, grayish structure 6 mm long, shaped like a pine cone, from which it got its name, and weighs about 0.1 gram in humans.

Von Frisch around 1911, wondered what role the pineal in the brain has in extraocular photoreception and initiated a pioneering experimental study of the pineal in response to light. He illuminated the pineal region of the European minnow, *Phoxinus phoxinus,* which had been previously blinded by the removal of the lateral eyes, and observed that their normally darkened skin became pale upon illumination. Eakin (1973) studied the median eye of the lizard *Sceloporus* (Figure 14.2), and reviewed the evolutionary development of the pineal organ in animals.

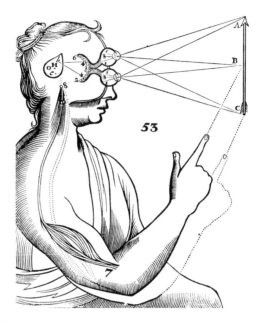

FIGURE 14.1 From René Descartes' *Tractus De Homine*, 1677. Descartes assigned the seat of the rational soul to the human pineal (H), in which the eyes perceived the events of the world and transmitted what they saw to the pineal by way of "strings" in the brain. (Photograph of wood engraving furnished by the Osler Library, McGill University, Montreal, Canada.)

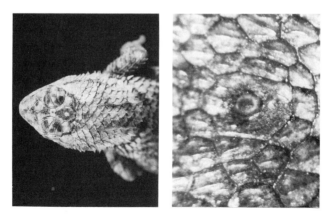

FIGURE 14.2 (a) *Sceloporus magister,* parietal eye; (b) *Anolis carolinensis,* parietal eye (dorsal). (Courtesy of Jan J. and E. Carol Roth.)

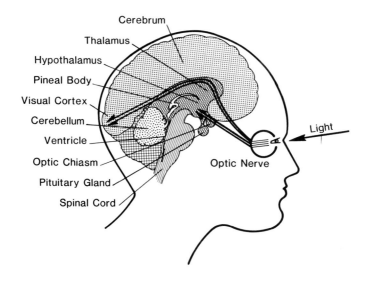

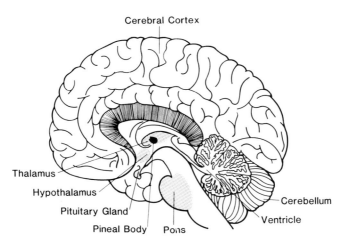

FIGURE 14.3   (a) Light through the eye that activates the pineal in the brain. (b) Diagram
of the brain's limbic system.

It was later found that if crushed pineal glands were introduced into water in
which tadpoles were swimming, the tadpole skin color bleached. Lerner et al.
(1958) were among the first to see a relationship between the extract of the pineal
glands and skin coloration. Lerner then isolated a substance from bovine pineal
glands and called it "melatonin," because it caused the contraction of the melanin
pigment granules. The action of melatonin on the skin brings about changes in skin
pigmentation in response to light and indicates that the pineal organ secretes the

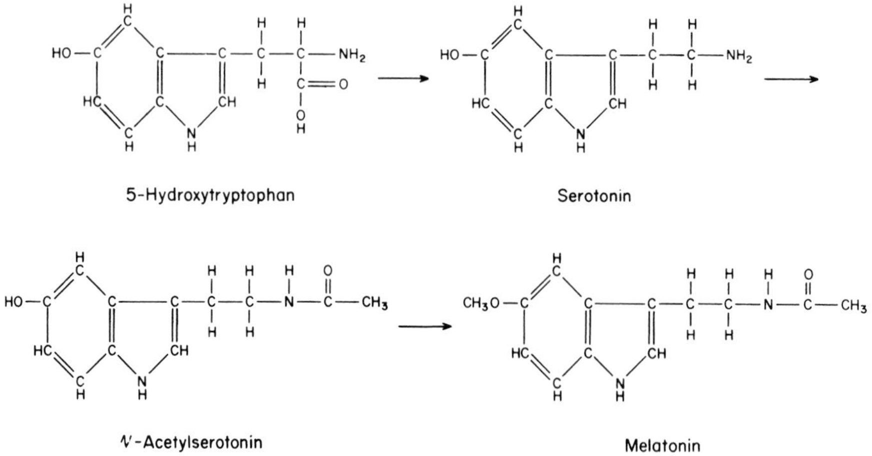

5-Hydroxytryptophan

Serotonin

N-Acetylserotonin

Melatonin

FIGURE 14.4   Synthesis of serotonin and melatonin.

chemical substance melatonin. It was then proven that melatonin was a hormone, that is, an active substance produced by the pineal. The melatonin molecule was found to be the indole amine, N-acetyl-5-methoxy-tryptamine (Figure 14.4). The precursor chemical necessary for melatonin synthesis is serotonin. Serotonin is a relatively widespread molecule in nature, it is found in cephalopod molluscs, amphibians, and in the pineal of all vertebrates. Serotonin is also found in the retina, pigment epithelium, and choroid of vertebrate eyes (Welsh, 1964; Quay, 1986). Surprisingly, serotonin is also found in plants such as bananas, figs, and plums.

The synthesis of melatonin begins with the amino acid, 5-hydroxy-tryptophan. Enzymatic action removes the carboxyl group (COOH). The product of this reaction is serotonin (Fig. 14.4). Another enzymatic reaction acetylates the molecules to form N-acetylserotonin which is then methylated to yield melatonin. The methylating enzyme, hydroxyindole-o-methyl-transferase, is found only in the pineal of mammals. The concentration of serotonin in the pineal is highest in bright light and during daylight or continuous light, while melatonin production falls abruptly in bright light and decreases at night or in darkness (Figure 14.5). The melatonin

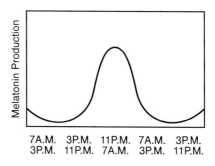

FIGURE 14.5   Melatonin Synthesis in bright light and in darkness.

rhythm exhibits 24-hour periodicity, which is directly related to circadian rhythms. The ability of melatonin to modify gonadal function suggests that its secretion has to do with the timing of the estrus and menstrual cycles of the reproductive processes. The activation of this pathway is controlled through a regular oscillating circadian rhythm. The elucidation of these mechanisms is due to the research of Bargmann (1943), Wurtman and Axelrod (1965), and Axelrod (1974). Their experimental findings indicated that the pineal is an extraocular photoreceptor system regulated by light and that it functions as a sensitive neuroendocrine transducer, a light-activated "biological clock." The molecule responsible for this activation is melatonin.

It is of interest to note that the neurotransmitter molecules acetylcholine and serotonin appear to play a role in photoperiodic phenomena. Neurotransmitter molecules are present in primitive organisms; for example, acetylcholine, epinephrine, and norepinephrine are found in protozoa and invertebrate nervous systems (Lentz, 1968). This finding suggests that the origin of photoperiodic behavior in animals may be related to a neurotransmitter molecule associated with photoreception. The apparent universality of these molecules in nervous systems suggests that a simple modification within the cell allowed for the coupling of light excitation with conduction, thus permitting transmission to occur.

This brings us to inquire: What is the structure of the pineal photoreceptor cells? Microscopy and electron microscopy reveal a striking observation: that there are highly differentiated photoreceptor cells in the pineal. These photoreceptor cells are similar in structure to the OS of retinal cones in the retinas of vertebrate eyes (Eakin, 1965; Kelly, 1965, 1971). Therefore, the pineal photoreceptors are analogous in structure to the retinal photoreceptors in the eye; for example, the absorption spectrum for the frog pineal is around 560 nm, comparable to the cone visual pigment iodopsin. Therefore, the pineal photoreceptor structure and photoreceptor pigment molecule are similar to that of the retinal photoreceptors in animal eyes.

An increasing body of experimental evidence has now become available on the pineal organ's structure and function (Ehrenkranz, 1983; O'Brien and Klein, 1986; Tamarkin et al., 1985; Wurtman, 1975). The pineal is, then, an extraocular photoreceptor that shares photoreceptor structures, photoreceptor pigment molecules, and biochemical processes similar to that of the retina of the eye. It functions as a light detector by measuring the light levels in the environment and is responsible for biorhythmic behavior. Therefore, it functions both as a photosensory organ and in the neuroendocrine system.

CONCLUDING REMARKS

Extraocular photoreceptors are found throughout the bodies of eyeless and blind animals. They function as light detectors to inform the animal of the presence of light and to measure light intensity as well as having other functions that activate rhythmic behavior, as in circadian reproductive rhythms.

The photoreceptor regions for extraocular photoreception are located in the skin, photosensitive neurons within specific ganglion cells, and in the pineal organ of the brain. Other light-sensitive regions of the brain are found in the hypothalamus, pituitary, and rhinocephalon. In the mouse and in the rat brain, the Hardian gland is the photoreceptor site that influences circadian rhythms.

The pineal gland synthesizes the hormone melatonin that affects pigmentation and reproductive processes and controls rhythmic behavior. The pineal is a sensitive neuroendocrine transducer which is activated by light. It is a "biological clock" that regulates both physiological processes and behavior.

The pineal photoreceptors are structurally similar to the retinal photoreceptors of vertebrate eyes, and the photoreceptor pigment molecule, like that in the retina, is the visual pigment rhodopsin. Although rhodopsin is most likely the photoreceptor molecule, other pigments have been identified with extraocular photoreception. Probably more than one pigment molecule participates in the photoprocess of extraocular photoreception.

How did the extraocular system evolve? The extraocular photoreceptor system evolved independently and continued to develop via the nervous system and the brain without being lost as the visual system evolved. In vertebrates, extraocular receptors are associated with older brain structures, such as the rhinencephalon and the pineal. This suggests a long history for the development and integration of extraocular photoreception in the vertebrate brain and that it continues to function in conjunction with the visual system. The effects of light on the mammalian pineal organ are mediated by a multisynaptic pathway that differs from the nervous impulses responsible for vision (Wurtman, 1975).

In comparing extraocular photoreception with the visual system, the threshold of light intensity necessary to produce a behavioral response is much lower than for vertebrate vision but is comparable to the visual threshold in invertebrates. The extraocular system has continued to function alone as evidenced in circadian rhythms, hormonal changes, and sexual cycles, but it also functions in conjunction with the visual system.

The effect of extraocular photoreception on behavior of animals, including human behavior, is truly extensive as indicated in reviews by Menaker (1976, 1977), Wolken (1988), Wolken and Mogus (1979, 1981), and Yoshida (1979).

Many interesting behavioral response mechanisms need to be understood. These include the light-mediated links of the pineal to seasonal disorders and even to states of mental health, the importance of the animal's skin to the endocrine system, and whether extraocular photoreceptors can detect polarized light and if so whether animals can perceive the plane of polarization via their extraocular photoreceptors for orientation, all of which need to be investigated further.

CHAPTER FIFTEEN

Bioengineering: Biomimetics

The whole discussion takes a new turn, however, when we consider that in such attempts at synthesis there is no need simply to follow nature's way exactly, or to use the same materials, which take millions of years to fashion by natural selection. Man's way is to find other materials and by short cuts produce what he calls "machines" that do the work more easily for him.

—J. Z. YOUNG, 1964, *A Model of the Brain*

Can the present information we have about photoreceptors and the optics of imaging eyes be exploited to develop imaging optic devices that can be applied to technology that will provide for human needs? Some examples of this developing technology include solar energy collectors, photoreceptor systems for (light) energy conversion, information systems, photochemical molecular computers, and optical imaging systems as prostheses for the visually impaired. These goals have stimulated researchers to develop devices and systems that mimic to some extent the way living organisms have engineered their photoreceptor, photosensory, and optical systems.

UTILIZING SOLAR ENERGY

The Sun is the major energy source for life on Earth; as a byproduct of natural degradative processes, over eons of time the Sun has indirectly provided us with fossil fuels as an energy source. As a result, we have been burning and depleting fossil fuels at an enormous rate to drive our technology. The extensive use of fossil fuels is polluting our atmosphere, is destructive to our environment, and is hazardous to life.

In order to decrease the use of burning fossil fuels as an energy source, alternative energy sources are being sought. The availability of solar energy on Earth has led engineers to develop the technology of solar panels for heating water to heat our homes and solar cells to drive engines. Photovoltaic solar cells directly convert light to electrical energy (Zweibel, 1990). These developments of solar devices are finding numerous applications; however, it has not diminished the need to develop more extensively solar radiation as an energy source.

To utilize solar energy more efficiently is to re-examine nature's process of photosynthesis which directly converts solar energy to chemical energy. Photosynthesis is the most efficient system for quantum energy conversion and storage of solar energy (Calvin, 1983). Ever since scientists began to probe how plants convert solar energy to chemical energy, they have wanted to replicate this process of energy conversion outside living plant cells. The development of solar energy conversion systems with an efficiency comparable to plants has not been accomplished and remains a great challenge to scientists. However, research is being done on the mechanisms of photosynthesis, that is, how the chloroplast is molecularly structured, like an energy conversion system to efficiently transduce light to chemical and electrical energy. These experimental findings are providing us with new insights on how this may be accomplished.

EXPERIMENTAL PHOTOSYNTHETIC SYSTEMS FOR ENERGY CONVERSION

> *Life, in its choices of molecules to do photochemistry with, has been extremely conservative and unoriginal. It has taken odd molecules lying around and used them with utmost skill to construct devices of high specificity, reliability, efficiency and sensibility. Under these circumstances it may turn out that scientists now, knowing so much more about photochemistry than Nature ever dreamed of, can construct devices, based on new molecules, that outdo the feats of organisms.*
>
> —M. DELBRÜCK, 1976, *Carlsberg Research Communication*

There are several experimental systems that attempt to replicate the process of photosynthesis outside the living cell by converting solar energy to chemical energy.

Experimentally, the simplest approach to creating an artificial photosynthetic system is to mix the naturally occurring chlorophyll, β-carotene with enzymes that serve as electron donors and acceptors. Other photochemical assemblies replace chlorophyll with porphyrins or photosensitive dyes and enzymes to bring about photo-oxidation reactions when excited by light. These photoreactions are carried out in solution or in lipid bilayer membranes, and in polymer films. A number of

analogous experiments are being pursued to find a photochemical system to replicate the photosynthetic process of energy transduction, in which a gradient of
chemical potential energy that is generated via photochemical oxidation and reduction reactions is measurable.

Other experimenters have turned to isolating the photoactive fractions from the
chloroplasts and to separating the antennae fractions that contain Photosystems I
and II (Figure 5.5). These fractions contain chlorophylls, carotenoids, and quinones in solution. Upon light absorption, a gradient of chemical potential energy
can be generated which results in measurable photochemical oxidation and reduction reactions, as in the photosynthetic electron transfer scheme as devised by Gust
and Moore, 1991.

Although these photochemical systems indicate that measurable energy transfer
has occurred, their quantum efficiency was very low when compared to photosynthetic plant cells. Therefore, these photochemical systems have drawbacks as
model systems. The question remains: Can the photosynthetic system be replicated
outside the living cell?

We have been exploring several other ways that an experimental photochemical
system may be achieved.

Chloroplastin

An experimental photoactive chemical system of photosynthetic cells can be reconstituted from the active components of chloroplasts. To isolate the photoactive
system from the chloroplast requires that it be solubilized. Chloroplasts are not
soluble in aqueous media, but molecular dispersion of the chloroplast can be
obtained by extraction with surfactants.

Surfactants are amphiphiles, possessing in the chemical structure of their molecules an ionic group that is water-soluble and an organic part that is water-
insoluble. When surfactants dispersed in water form micelles of concentric
lamellae, these lamellar structures consist of interfaces, like those of bilayer lipid
photoreceptor membranes.

Digitonin, a digitalis glycoside, is a non-ionic surfactant whose chemical structure resembles cholesterol (Figure 15.1a,b). Digitonin (1%–2%) dispersed in water has a strong attraction for complex molecules, particularly lipids and naturally
occurring pigments, such as carotenoids, chlorophylls, and the visual pigment
rhodopsin.

The role of the digitonin micelle is to react with one of the substrates while
simultaneously attracting the other substrate to the same vicinity. This parallels the
behavior of an enzyme in bringing the reactants together. The digitonin micelles
can then be used for many biological assemblies since the interactions responsible
for micelle stability are similar to those which stabilize bioaggregates. These
supermolecular assemblies compartmentalize reacting molecules and have a pro-
nounced catalytic effect on energy and electron transfer by virtue of the potential
gradients at the lamellar interfaces.

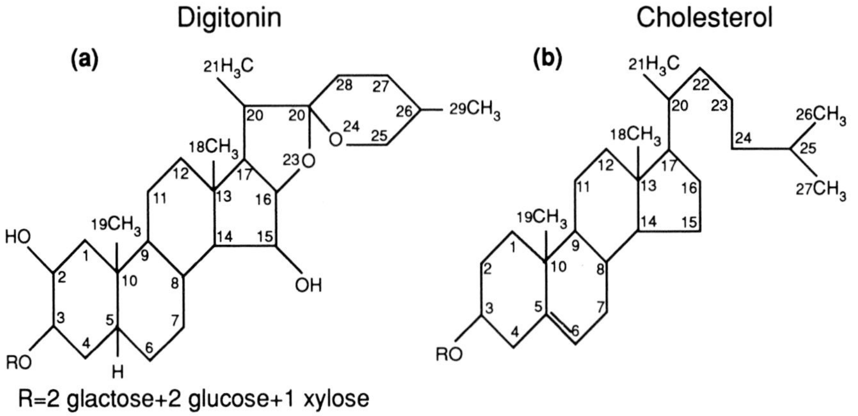

FIGURE 15.1   Chemical structure of digitonin (a) compared to that of cholesterol (b).

The chloroplast extracted with 1%–2% digitonin (or other surfactants) is a molecular dispersion of the chlorophyll complex. The extracted chlorophyll complex can be further separated by high-speed ultracentrifugation, after which a clear green fraction is obtained in the centrifuge tube. This isolated green fraction is referred to as *chloroplastin*.

Chloroplastin is birefrigent when observed through crossed polarizers; hence, there is an alignment of chlorophyll molecules in the complex. This is structurally observed when a drop of chloroplastin is evaporated from the surface of a glass slide and lamellar rings are formed. When this lamellar structure is scanned with a microspectrometer at 675 nm (the major absorption peak for chlorophyll), chlorophyll is found to be concentrated and aligned in the lamellae and not in the interspaces, thus mimicking the chloroplast molecular structure (Brown and Wolken, 1979; Wolken, 1986). What is most interesting is that chloroplastin is photoactive and can perform some of the photometabolic processes associated with photosynthesis. Upon light absorption, it will photoreduce chlorophyll, evolve oxygen, and, in the presence of the right cofactors, perform some of the primary steps of photosynthesis such as turning an inorganic phosphate to an organic phosphate—ATP, the driving energy source for all living systems (Serebrovskaya, 1971; Wolken, 1975, 1986). Therefore, chloroplastin provides an experimental model system for the study of light energy conversion to chemical energy, as does the chloroplast photosynthetic system.

Chlorophyll in a liquid crystalline system

A photochemical experimental system which has analogies to chloroplastin can be assembled with chlorophyll (or other photosensitive dyes) in a cholesteric liquid crystal. Cholesteric liquid crystals have similarities to cholesterol in chemical

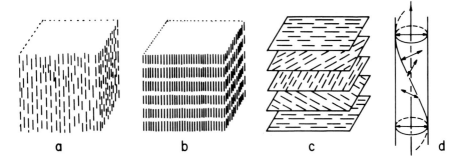

FIGURE 15.2   The molecular structure of three main types of liquid crystals: a) Nematic, the elongated molecules are randomly distributed. b) Smectic, the molecules are ordered in layers (lamellae) of equal thickness. c) Cholesteric, the molecules rotate regularly from plane to plane. d) Helix of a cholesteric rod. (From Wolken, 1984, and Brown and Wolken, 1979.)

structure (Figure 15.1b). The molecular structure of cholesteric nematic liquid crystals is described in Brown and Wolken (1979) and Wolken (1984). An important characteristic of liquid crystals is that they possess structural order (Figure 15.2). In a liquid crystalline system, the molecules are uniformly aligned along their long axes in the same direction, much like the chlorophyll molecule in chloroplasts (Figure 15.3). Chlorophyll molecules in a nematic liquid crystal perturb the regular arrangement of liquid crystals and influence their dielectric properties. Chlorophyll molecules will orient in liquid crystals (Journeaux and Viovy, 1978), and a chlorophyll-lipid-protein can be reconstructed into a liquid crystalline system (Ke and Vernon, 1971; Wolken, 1975).

These few experimental models for light transduction indicate that a photo-

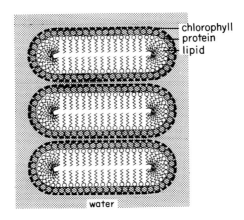

FIGURE 15.3   Model of chloroplastin micelles.

chemical multicomponent system can be assembled that converts light to chemical and electrical energy but that the quantum efficiency so far is not comparable to plant photosynthesis.

When the biophysics and biochemistry of phototransduction mechanisms are better understood scientists will be better able to replicate this photoprocess with physical chemical systems outside the living plant cells.

INFORMATION PROCESSING SYSTEMS

The kinds of information that are important to living organisms are (1) genetic information, which does not get feedback from the organism but is passed on from generation to generation; (2) the sensory information, which has considerable feedback into the storage system but is not passed on from generation to generation; and (3) the communicated information, which does have feedback and is passed to the next generation. It is the sensory and communicated information that concerns us.

Living organisms, from bacteria to humans, respond to physical stimuli (light, sound, pressure) via their receptors. Receptors detect and measure the strength of these various stimuli, and this information is communicated to the organism which determines behavior.

In animals, the primary receptors are visual, auditory, and tactile, and the processing of this information is a function of the nervous system. Upon absorption of the stimulus, the light energy is transduced into chemical and electrical signals. These electrical signals are further processed and transmitted through the nervous system's axons, neurons, and synapses to arrive, completely transformed, in the cerebral cortex of the brain.

The maximum likely storage capacity of the human brain, which has on the order of 10^{10} nerve cells in the cerebral cortex, is around 10^{10} to 10^{11} bits of information. Since the sum sensory inflow from all the sensory organs to the nervous system has been estimated at 10^7 bits per second; in a lifetime of 10^9 seconds, the total would be 10^{16} bits of information, which is considerable (Young, 1971).

How the eye receives information from visual pigments and processes visual information that encodes essential details about shape, pattern, and color to the visual cortex of the brain is very complex (Lythgoe and Partridge, 1989). Studies of the neural visual processing were undertaken by Wiener (1964) and McCulloch (1965). They based their analysis on the available information on neural networks, tracing the neural circuitry from the retina and theorizing the pathways that led from the retina to the visual cortex in the brain. These earlier studies were followed up by Hubel and Wiesel (1979) who discovered how the visual system processes information, that is, the neural network circuitry to specific areas in the visual cortex of the brain. This has greatly enhanced our knowledge of the visual process, and for it they were awarded the Nobel Prize in 1981. However, how the eye processes visual information is not completely understood. The complexity of visual processing is discussed by Marr (1982). How the eye processes information remains a major task for investigators of neural and visual science.

Is the photoreceptor cell a molecular computer?

A computer, by simple definition, is any system that receives information from the environment, is altered by this input, and then puts out information. Therefore, any system which has large amounts of incoming and outgoing information relative to its internal storage capacity can be considered a computer.

We can think of the visual photoreceptor cells in the retina as a photomolecular computer. The visual photoreceptors are structured of stacked membranes. The membranes are bimolecular lipid-protein layers of around 100 Å in thickness, comprising proteins, enzymes, and photoreceptor pigment molecules. There are 10^6 to 10^9 photoreceptor pigment molecules within the photoreceptor structure, arranged as monolayers on all the membrane surfaces for maximimum light absorption. The cross-sectional area of each photoreceptor molecule on the membrane surface is estimated to be about 50 $Å^2$. This approximates the cross-section of the rhodopsin molecule in the retinal rods. Kühn (1968) calculated that it should be possible to store a single bit on an area of 100 $Å^2$, so each photoreceptor molecule can receive a significant amount of information for processing visual information.

In the retinal rod, photosensitive rhodopsin molecules are associated with all the membrane surfaces. Upon light absorption at a particular wavelength, rhodopsin changes in the pigment chemical geometry occur (e.g., retinal from *cis* to *trans*). In this process, conformational changes also occur in the protein to which retinal is complexed. Light then triggers the system and the rod elongates and recovers in the darkness to its original state to be triggered again by light (Figure 15.4).

Certainly, the visual photoreceptor cell can be considered an analog of a photochemical molecular computer. A simple model for light reception, storage, and recall can be conceptualized as schematically illustrated in Figure 15.4. In this retinal rod photochemical model, information is received by the absorption of a particular wavelength of light, erased by another wavelength of light, or restored by darkness to its original state to be triggered again by light. In the photocycle, the elongation and retraction of the retinal rod structure is like a spring, or a jack-in-the-box. In a sense, this photoreceptor molecular model behaves like a computer that function through photochemical, rather than electronic, changes. Light induces a switching similar to that which occurs in photovoltaic silicon microchips. The photoreceptor's behavior parallels the function of a flip-flop in a digital computer. A flip-flop always has one of two values, 0 or 1, but when an electrical signal reaches it, it switches its current value (i.e., if it was 0 it becomes 1, or vice versa). This is very widely used in both input and storage circuits.

So, to make a photochemical computer, one simply needs a way to use light

Photoreceptor

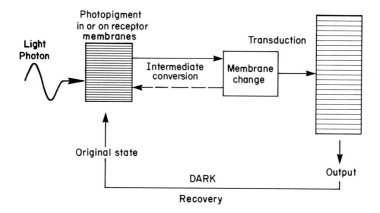

FIGURE 15.4   Schematic for a conformational change of the photopigment in the photo-receptor membranes upon photoexcitation.

switching, that is, to let a light signal at one wavelength control a second signal at another wavelength. This requires one or more photosensitive pigment molecules. The naturally occurring photoreceptor pigment molecules are carotenoids, rhodopsins, bacteriorhodopsin, and phytochrome. These pigments are conjugated organic molecules, composed of single and double alternating chemical bonds—polyene chains that have large nonlinear coefficients.

Nature has selected these pigments molecules for photoreception. For example, the visual pigment chromophore is retinal, whose spectral absorption peak is around 370 nm, and when complexed with its protein opsin to form rhodopsin, the color changes from yellow to reddish-purple and its absorption is shifted to around 500 nm. Light bleaches rhodopsin to release retinal from opsin, which in darkness will recombine to form rhodopsin again. Bacteriorhodopsin, like rhodopsin, has its absorption maximum peak around 570 nm, whereas the intermediate of the photocycle absorption is shifted to around 412 nm in the blue (Lanyi, 1992). Although the bacteriorhodopsin system is restricted to 570 nm in the yellow, the entire visible spectrum (green, yellow, and red light) is usable. These spectral differences between the basic and the intermediate states can record information in a way similar to the light–dark distribution in a photographic film. Phytochrome is another photoreceptor pigment molecule that has similarities to rhodopsin and switches its absorption peak from the red-absorbing 660 nm to the far-red-absorbing 730 nm. There are other naturally occurring photochromic organic pigment molecules, for example, chromones (benzyl-γ-pyrone), pyrans, and flavones (2-phenyl chromone), which are widespread in plants and animals. Photochromic pigment molecules change color upon light absorption, and the color changes are wavelength-dependent and are reversible. To create a photoreceptor membrane, these pigment molecules can be embedded in polymer films that have high optical transparency. The photoreceptor pigment molecules in the membrane can receive a light signal at its absorbing wavelength, store or transfer the information, and be ready to receive a second light signal at another wavelength of light, and so on. The stored spectral information can be recalled by the original wavelengths of light. Once the information is received or transferred, it can be erased, and the photoreceptor molecules restored to their original chemical state are ready to function again.

The development of a photomolecular computer, based on the light switching of photoreceptor pigment molecules, would have enormous computing ability, representing an advantage over the present silicon circuit microchips, which are two-dimensional structures. Photoreceptors of living cells are three-dimensional structures, and a three-dimensional structure would facilitate the development of parallel processing, now the goal of computer scientists. This would overcome the limits of present computers by providing more information capacity. However, before a biophotomolecular computer can be developed, many problems must be solved, such as the necessary optical and photochemical hardware arranged in a neural architecture that would produce a practical computer capable of dealing with random problems.

MODELING OPTICAL IMAGING DEVICES

Optical engineers, in developing "machine vision," have long wanted to design an imaging system that functions like the "seeing" eye of animals. To replicate the visual system of an eye is a difficult, if not impossible, task. However, imaging systems have been developed with light sensors, photovoltaic photocells, and silicon microchips which, together with television cameras, can acquire visual information about the environment. These photo-chemo-electro devices have been adapted to robots and are also being explored as aids for the visually impaired.

Much can be learned from studies of the optical and photoreceptor systems of animal eyes and applied to the design of new imaging devices. Nature has more than anticipated the development of modern optics. In fact, every known type of optical imaging system can be found among invertebrates, from pinhole to camera-type eyes to compound eyes and to refracting-type eyes. Among these animals are found eyes with prismatic corneas and variously shaped lenses, some animals having evolved eyes with reflecting surfaces (mirrors) or fiber optic light wave guides.

Lenses are spherically or aspherically shaped, thus providing many curved surfaces for light reflection and refraction. The lenses are formed of layers (interfaces) with varying indices of refraction, in which the index of refraction is highest in the center of the lens and decreases toward the periphery.

Eyes are highly sophisticated optical and photochemical systems, designed to improve the ability of animals to visualize their world.

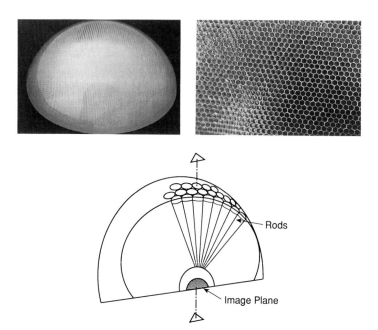

FIGURE 15.5   Fiber optic filament formed into a hemispheric lens. An optical imaging device that can result in a wide-angle view without the distortion inherent in conventional wide-angle lenses.

The structure and optics of the compound eyes of arthropods, described in Chapter 12, are of special interest in the design of imaging devices. Compound eyes are composed of eye facets called ommatidia. Each ommatidium is an eye whose optical system consists of a corneal lens (L_1) and a crystalline cone (L_2). In some arthropods, the corneal lens and crystalline cone form a single lens, while in others there are double lenses and even triple lenses. Depending on the spatial relationship of the lenses (L_1 and L_2) to the photoreceptors, the eye structure will give rise to either an apposition or a superposition image (Figure 12.2).

There are several types of existing optical lens systems that have been modeled after invertebrate imaging eyes. The most common optical system is that resembling a simple eye that uses a single aperture. A more complex optical system is a multi-aperture optical system modeled on the compound eye structure. A multi-aperture lens can be modeled using hundreds of fiber optic filaments shaped into a hemisphere (Figure 15.5). This modeled lens approximates a compound eye, in which each fiber optic filament represents an eye facet. Light passes down each fiber optic filament much as it does down an ommatidium of a compound eye. This compound lens can focus an image that can then be projected on photographic film or adapted to imaging devices.

A multi-aperture lens array with glass rods arranged in a hemisphere was developed by Zinter (1987). In his model lens, each rod (treated to create a graded index of refraction) acts as a single lens that allows it to transmit a small section of the scene. At the focal point, each rod produces an overlap, resulting in a continuous image. Images obtained from each of these rods were transformed via the fiber optic bundles and then superimposed creating an intensified image. This optical array gives a superposition image much like that of a compound eye. Such an optical device has a wide field of view and can operate at relatively low light levels. This type of lens would be useful in detecting moving objects and for discerning their shapes.

Another novel imaging lens and optical scope was modeled after the eye of *Copilia quadrata,* a deep-sea crustacean whose eye structure and optical system were described in Chapter 12. This optical system has two lenses: an anterior, biconvex lens (L_1) and, some distance away, a posterior, pear-shaped lens (L_2)—a telescopic eye. How the *Copilia* optical system functions for imaging was illustrated in Figure 12.21. In this optical system, the lens, L_1, forms an image at I_1, which is intercepted by lens L_2 and imaged at I_2. The effect of the second lens, L_2, is to condense the partially focused images from L_1 onto a much smaller area, thereby increasing the image brightness and acting as a light amplifier.

Using the anatomical structural data of the *Copilia* eye (the shape and spacing of the lenses L_1 and L_2), the optical system was reconstructed. In developing the optical system, it was found that the curvatures of the front and back surfaces of the lenses L_1 and L_2 must obey the relationship indicated in Figure 15.6. The radius of curvature of lens L_1 is such that for the front surface $R = r$ and for the back surface $R = 2r$. These geometric ratios are critical to its function as an optical system. The L_2 lens is actually two hemispheres, with the ratio of their radii being 2:1, joined by a section of a cone whose sides slope from $16°$ to $19°$.

The size of the lens L_2 is dependent on its end use; for it can be very small or very large, providing the geometry, i.e., the curvature of the lens and angle it subtends, is maintained at the calculated values. The lens L_2 can function as a light-concentrating imaging lens and can be used without any additional lenses. To give the scope a greater field of view, a positive meniscus lens can replace the lens L_1 thus correcting for spherical and chromatic aberration. A graded index can be incorporated in the core of the lens to improve its imaging capability (Figure 15.6). These lenses can be made of acrylics, other ophthalmic polymers, or glass.

To test the optical system for imaging each lens focal point was determined and the lenses mounted in a tubular polyethylene housing, designed to fit the body of a (Leica and Canon) camera. The optical system was adjusted in the camera so that it would focus an image on the plane of the film, and images on the film could be obtained from ten inches to infinity. The camera was subjected to photographic tests under different environmental constraints (e.g., air, water, murky water, sea water). Panatomic X infrared and standard black and white and color films were used. Since the objective was to detect the limits of the system in terms of lighting

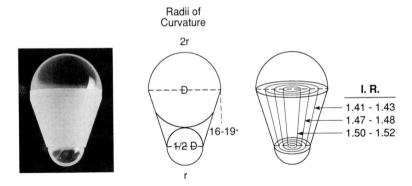

FIGURE 15.6   A light-concentrating lens (modeled after the *Copilia,* crystalline lens, L_2).

and resolution, moving as well as static objects were photographed. As an example, photographic images obtained with indoor lighting using Panatomic X (ISO 32) at shutter speeds of 1/1000 and 1/5000 of a second were sharp. As the ISO number increased, the minimum shutter speed for overexposure of an image increased, due to the fact that the lens L_2 is a light concentrator. The resulting images are sharp due to high contrast.

This novel lens, L_2, and scope (L_1 and L_2) has many advantages over a conventional optical system, especially where there is a need to resolve images at relatively low light levels and to track moving objects. The importance of this optical system is that it greatly facilitates scanning and provides for a high-aperture and high-resolution optical device. Adaptations of the scope (Figure 15.7) as a prosthesis for the visually handicapped indicate that it has considerable potential as an aid for resolving images in low light levels of illumination and for observing moving targets (Wolken and Mogus, 1988). Other applications for the lens system include its use for surgical scopes, microscopy, optical scanners, navigational devices, astronomical cameras, gun sights, and solar energy collectors (Wolken, 1987, 1991).

The lens, L_2, is a non-imaging (CPC) lens and can be adapted to solar energy collecting devices. Winston (1975) designed a similar lens. Welford and Winston

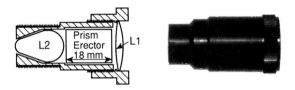

FIGURE 15.7   An optical telescope (modeled and constructed after the optical system of the *Copilia* eye).

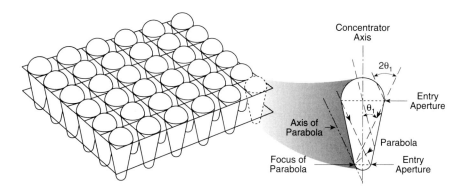

FIGURE 15.8   An array of lenses, L_2, used for solar energy collection.

(1989) assembled a series of these lenses in tandem. They used a mirror to focus the solar light on these lenses to maximize solar energy collection and storage (Figure 15.8).

These experimental models for energy conversion, informational systems, and optical imaging devices indicate that the basic features of living photoreceptors and optical systems can be replicated in time. However, much more research of natural systems is needed for their full development.

APPENDIX

Optics: How Images Are Formed

A general review of the basic principles of the physical and geometric optics involved in the formation of images by lenses will be helpful in understanding the optics of the various eyes and visual systems that were discussed.

Those interested in the mathematical derivations and formulations in the design of imaging lenses should refer to an optics text (Kingslake, 1983; Wood, 1988; and to others that are cited in the References).

HOW IMAGES ARE FORMED

Optical devices that redirect the paths of light rays to form an image are lenses, mirrors, and prisms. An imaging system may contain any combination of these optical devices. A lens is essentially a curved surface or a series of curved surfaces which differ in index of refraction from that of the surrounding medium. A lens redirects the path of light in such a way as to form a focused image. Convex lenses will cause a beam of light to converge and concave lenses will cause a beam of light to diverge (Figures A.1 and A.2). When light passes through a converging lens, the refracting light rays form a focused real image. If the light rays diverge after passing through a lens, a virtual image is formed. A real image can be projected on a screen, a virtual image cannot be made visible on screen, but can be viewed directly through the lens.

The direction of light also changes when light enters matter. This effect is known as refraction. The amount of refraction of light can be determined from the index of refraction of the material. The index of refraction n is defined as,

$$n = \frac{c}{v} > 1$$

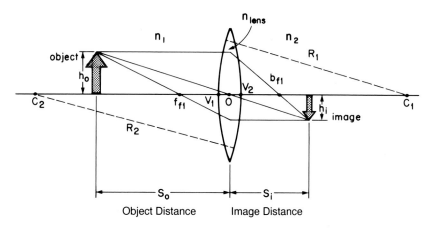

FIGURE A.1 A positive (convex) lens with index n_{lens} focuses a real image of the arrow on the opposite side of the lens. The image distance is therefore positive. Shown are the points representing the back focal length (b_{fl}) and front focal length (f_{fl}), object and image distance (S_o) and (S_i). Note that the b_{fl} and f_{fl} are equal because the radii of curvature are the same for both sides. The image is smaller than the object. This is due to the position of the object, which is just outside the f_{fl}.

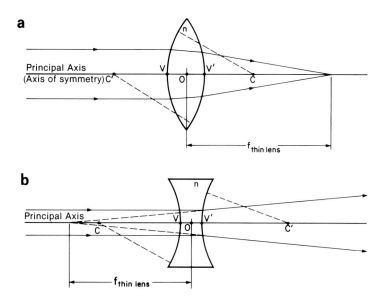

FIGURE A.2 (a,b) Convex and concave lenses, respectively. C denotes the center of curvature of a lens, and dotted lines emanating from this point denote radii of curvature. V denotes a vertex, and O is the optical center of the lens. The focal length f of a lens is defined as the distance from the lens at which the lens converges the entering parallel rays.

where c = 3×10^8 m/sec. is the speed of light in a vacuum, and v is the speed of light in the medium.

REFLECTION-REFRACTION

When a beam of light strikes a surface, its intensity is usually split between the reflected and refracted beams. The amount of the beam that is reflected or re-fracted (transmitted) is governed by the reflection (R), transmission (T), and absorption (A) coefficients, where R + T + A = 1. The coefficients R and T can be calculated from the well-known Fresnel formula.

For the reflected beam, the angle of incidence equals the angle of reflection. These angles can only range from 0° to 90°. Snell's law defines the degree of refraction, or bending toward or away from the normal to the surface, for the transmitted beam. Snell's law is given as: $n_1\sin\theta_1 = n_2\sin\theta_2$, where n_1 is the refractive index of the first medium, n_2 is the refractive index of the second medium θ_1 is the angle of the incidence, and θ_2 is the angle of refraction (Figure A.3). As the value of θ increases from 0° to 90°, the sine function increases monotomically from 0 to 1. No real angle can have a sine θ greater than 1. When light is propagating in a material of high index of refraction (glass, n = 1.52; water, n = 1.33) and is transmitted through a surface into a medium of lower index of refraction, then the angles may be such that Snell's law gives $n_1/n_2 \sin \theta_1 \geq 1$. Since this condition cannot occur, the result is that light will not escape the medium in which it is traveling and will be bent back into this medium. This phenomenon is called *total internal reflection* and the angle at which this occurs is called the *critical angle* for that interface. This angle may be determined by rewriting Snell's law for the case where sin θ_2 = 1:

$$\sin\theta_c = \frac{n_2}{n_1}$$

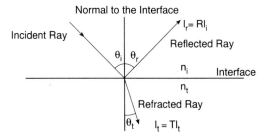

FIGURE A.3 Diagram of reflection and refraction at an interface. I_i is the intensity of the incident light. R and T are the reflectance and transmittance, where R + T = 1, ignoring any absorption. n is the index of refraction of the substance, θ is the angle that a light ray makes with the normal to the surface, and subscripts i, r, t, denote incident, reflected, and transmitted quantities, respectively. Note that $\theta_i = \theta_r$.

where n_1 is the refractive index of the medium in which light is traveling and n_2 the refractive index of the medium into which light is exiting.

THE FOCAL LENGTH OF A LENS

The focal length of a lens, f, is defined as the optical path length from the center of the lens to the point at which a parallel beam of light converges, or the optical path length from the center of the lens to a point from which the ray seems to diverge, as in Figure A.4. The focal length is positive if the parallel rays converge on the side of the lens opposite to which they are incident. The focal length is negative if the rays appear to diverge from a point on the same side of the lens on which they are incident. Lenses with positive focal length (positive lenses) produce real images and lenses with negative focal length (negative lenses) produce virtual images.

The focal length of a lens can be calculated from the radius of curvature of the lens surface, the index of refraction of the lens, n, and the index of refraction of the lens environment, n_2. Accordingly, the focal length of a lens is calculated from the following equation:

$$\frac{1}{f} = \frac{n_{1\ lens} - n_1}{r_1} + \frac{n_{1\ lens} - n_2}{r_2}$$

where r_1 is the radius of curvature of a surface. For most cases $n_1 = n_2$ and the equation reduces to the following:

$$\frac{1}{f} = (n_{lens} - n_1)\left(\frac{1}{r_1} + \frac{1}{r_2}\right)$$

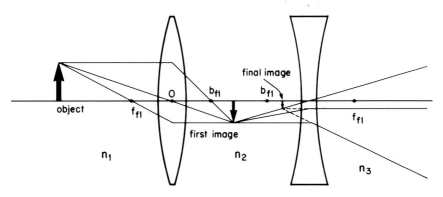

FIGURE A.4   The optical path length from an object through the center of the lens to a point from which the light ray diverges.

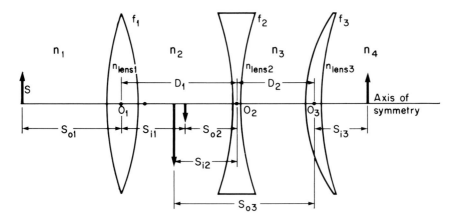

FIGURE A.5   A system of lenses with different focal lengths. Image positions can be calculated by successively applying the Gaussian Lens Equation if all focal lengths and distances between lenses are known.

where the medium is air, $n_1 = 1$. The sign ($\pm$) of the radii are such that for a biconvex lens r_1 is positive and r_2 is negative.

Using the variables in Figure A.5, the Gaussian Lens Equation relating object distance, image distance, and the focal length is given as:

$$\frac{n_1}{s_o} + \frac{n_2}{s_i} = \frac{n_{lens} - n_1}{r_1} + \frac{n_{lens} - n_2}{r_2}$$

The computed focal length, f, is defined as the optical length from the center of the lens to the point where parallel rays converge. One can find the distance from the center of the lens to the point of convergence of the parallel rays by setting $s_o = \infty$ and solving the Gaussian Lens Equation for s_i. In addition, let $n_1 = n_2$. This special case gives rise to the Thin Lens Equation which applies to thin lenses of the type found in optical devices such as telescopes. The Thin Lens Equation is:

$$\frac{1}{s_o} + \frac{1}{s_i} = \frac{1}{f}$$

The relationship between the distance from an object to the lens and the distance from the image to the lens is given by the above equation, where S_o is the object distance, S_i the image distance, and f the focal length.

MAGNIFICATION

Lenses alter the size of an imaged object. How object size has changed is known as the transverse magnification, M_t. Transverse magnification, M_t, is defined as the ratio of the image height to the object height:

$$M_t = \frac{h_i}{h_o}$$

The magnification may also be calculated from the image distance and the object distance as follows:

$$M_t = \frac{-n_i s_i}{n_2 s_o}$$

when $n_1 = n_2$, the magnification reduces to

$$M_t = -\frac{s_i}{s_o}$$

When the image distance is negative, the image is inverted, and when magnification is positive, the image is erect.

MULTIPLE LENSES

In an optical system of two or more lenses where all of the lenses are aligned on the same axis of symmetry, the image formed by one lens, whether real or virtual, serves as the object for the next lens (Figures A.4, A.5).

In the above analysis, it is assumed that the refractive index of the medium is equal around all sides of the lens. This is not the case for a biological lens system operating in air (n = 1.00) or water (n = 1.33). When the medium in front of the lens has a different refractive index from that behind the lens (e.g., the cornea and lens system of the vertebrate eye), the Gaussian Lens Equation cannot be simplified to determine the focal length. The ray path for such an optical system may be determined from mathematical analysis or from computer simulation. Computer simulation has simplified both lens analysis and lens design. However, the optical analysis of lens systems in eyes of living organisms where there are multiple lenses still needs to be experimentally determined.

LENS ABERRATIONS

Due to the geometry of spherical lenses, two principal defects occur. The first is *spherical aberration* which occurs in the case where the radius of a lens surface sweeps out more than 40° of arc. Light rays passing through the center of the lens are brought to focus at a greater distance than light rays at the margin of the lens. The marginal zone of the lens can be thought of as having a different focal length than the central zone of the lens. The result is a loss of sharpness in the image.

A second defect in lenses is known as *chromatic aberration*. If a beam of white light passes parallel to the optical axis of an ordinary lens, the shorter wavelength of the spectrum, blue light, will be brought to focus at a point nearer to the lens than the longer wavelength, red light. This failure of a lens to converge light of different wavelengths to the same point is due to the variation of the refractive index with wavelength.

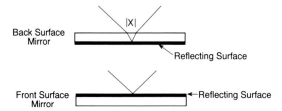

FIGURE A.6   The back surface mirror displaces the light beam before reflecting it, while the front surface mirror does not.

MIRRORS

Images can also be formed by mirrors. As a lens redirects the path of light rays by refraction, a mirror redirects the path of light rays by reflection. There are three types of mirrors commonly used in optics: flat or plane mirrors, spherical mirrors, and parabolic mirrors.

The flat or plane mirrors used in optics are slightly different from the average mirror used in the home or car. Most plane mirrors used in optics are front surface mirrors. The difference from the average mirror is that the reflecting surface is on the front (Figure A.6a,b). This is done to avoid the small angle dependent translation in the beam that would occur if a back surface mirror were used. The image in a flat mirror appears to be the same distance behind the mirror as the object in front of the mirror. The image is also of the same height, but is a back to front inversion in the mirror.

Spherical mirrors are much like spherical lenses only simpler since there is only one index of refraction instead of two or three. The focal point of a spherical mirror is where the light rays parallel to the optical axis appear to come to a point. This point is at $r/2$, where r is the radius of curvature, and is always halfway between the vertex and the center of the mirror. The focal length of a spherical mirror is considered positive if the focal point is on the same side of the mirror as the light rays and negative if the focal point is on the opposite side of the mirror from the light rays (Figure A.7). Due to a similarity in geometry, the spherical mirror equation is almost identical to the Gaussian Lens Equation except that no indices of refraction enter into it. For spherical mirrors we have the following:

$$\frac{1}{f_m} = \frac{1}{s_o} + \frac{1}{s_i}$$

The magnification is also similar to the Gaussian Lens Equation as follows:

$$M_t = -\frac{s_i}{s_o}$$

Spherical mirrors also suffer from spherical aberration but not from chromatic aberration. Spherical aberration can be corrected by using a parabolic mirror. The

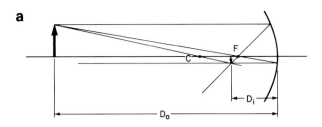

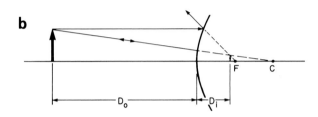

FIGURE A.7   (a) A ray tracing diagram for a concave (converging) mirror. (b) A ray tracing for convex (diverging) mirror. The focal point f is located at a distance r/2 from the vertex; C, center of the curvature.

surface of a parabolic mirror (Figure A.8) can be described by the equation $h(x) = ax^2$, where a and x are arbitrary constants. When the slope of the tangent to the surface is one, an incoming parallel ray is deflected exactly 90° through the focus and parallel to the horizontal (slope 1 is equivalent to a 40° angle with the horizontal). The slope of $h(x)$ is the first derivative $h'(x) = 2ax$ which equals 1 when $x = 1/2a$. Hence the focal point f of the parabolic mirror is at $h(x) = a(1/2a)^2 = 1/4a = f$.

A mirror with parabolic faces is designed to concentrate faint light and is an ideal light collector (Welford and Winston, 1989). Such an image collector (Figure 15.6) is an excellent mechanism for a biological optical system that does not depend upon refracting optics (Wolken, 1987).

PRISMS

A prism is a wedge that is generally 45°–90°–45°, oriented so that light enters and leaves normal to the hypotenuse side, as shown in Figure A.9. An infinite series of prisms can function as a convex lens, and like a convex lens it can cause the beam of light to converge and thereby function as an eyepiece.

An application of a prism in optical imaging systems is to reinvert an image. This is illustrated in Figure A.9a,b, where the letter R was selected as the object

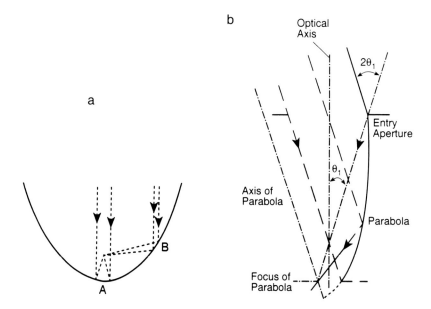

FIGURE A.8 (a) Image formation by a parabolic mirror shows that the image formed by the portion of the paraboloid at B is inclined at 90° to the image formed by the portion at A, and the image is perpendicular to the light rays (as schematized by Wood, 1988). (b) Image formed by a parabolic lens (after Welford and Winston, 1989).

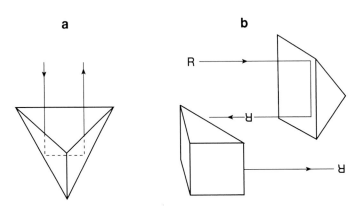

FIGURE A.9 (a) A prism is 45°–90°–45° oriented so that light enters and leaves normal to the hypotenuse. It is a corner reflector (retro-flector) and sends the light ray back in its original direction. (b) The letter "R" is used here as the object. The two refractions serve to invert the image R, as shown.

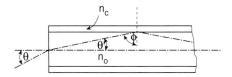

FIGURE A.10   Finding the numerical aperture of a fiber optic filament.

because it has no symmetry and no orientation or reversal can be distinguished. The first two reflections shown serve to turn the image upside down without exchanging left and right. The image of R is then mirror-backwards and therefore not yet inverted. Another prism oriented sideways as shown reverses the other dimension of the image and results in an inverted R.

FIBER OPTICS

Developments in fiber optics have taken advantage of optical principles known for many years (Wood, 1988). Fiber optics are drawn of long threads of glass (or polymer) fibers about 1–10 μm in diameter (the limiting diameters range from about 0.85 to 1.5 μm). These fibers are coated (clad) with an optically lower refractive index than that of glass or a jacket, to prevent the leakage of light. Having a circular cross-section, light is transmitted along the length by a process of multiple total internal reflection.

The numerical aperture of the fiber is the sine of the slope angle q of the steepest entering ray that is just at the point of total internal reflection inside the fiber. In Figure A.10 it is seen that if θ is the critical angle, $\sin\theta = n_c/n_o$, where n_o is the refractive index of the fiber and n_c that of the cladding material. The numerical aperture is:

$$NA_{fiber} = n_o \cos\theta_c = n_o \sqrt{\frac{n_o^2 - n_c^2}{n}} \text{ or } = \sqrt{n_o^2 - n_c^2}$$

Thus, if the index of the fiber is 1.55 and that of the cladding is 1.50, the numerical aperture will be 0.39.

A fiber optic can be a rod cylinder with the axis of symmetry as the densest region, and the optical density (index of refraction) decreases from the center to the edge continuously as $1/r^2$, where r is the radial distance from the center of the cylinder. The refractive index of the cylinder follows a parabolic index gradient, referred to as a graded index of refraction (GRIN) lens. A graded index rod lens concentrates the light along the optical axis and greatly increases the transmission of light. Fiber optic light guides are found in eyes of certain insects and crustaceans.

References

Able, K. P. and M. A. Able (1993) Daytime calibration of magnetic orientation in a migrating bird requires a view of skylight polarization. *Nature,* **364,** 523–525.

Abrahamson, E. W. and R. S. Fager (1973) The chemistry of vertebrate and invertebrate visual photoreceptors. *Curr. Top. Bioeng.,* 125–200. *Bioenergetics,* Vol. V, Academic Press, New York.

Adler, K. and D. H. Taylor (1973) Extraocular perception of polarized light by orienting salamanders. *J. Comp. Physiol.* **87,** 203–212.

Amesz, J. (1973) The function of plastoquinone in photosynthetic electron transport. *Biochim. Biophys. Acta,* **301,** 35.

Andersen, M. C. and A. M. Brown (1979) Photoresponses of a sensitive extroretinal photoreceptor in *Aplysia. J. Physiol.* **287,** 267–282.

Arnon, D. I. (1965) Ferrodoxin and photosynthesis. *Science,* **149,** 1460.

Austin, G., H. Yai, and M. Sato (1967) Calcium ion effects on *Aplysia* membrane ion potentials. In *Invertebrate Nervous System,* C. A. G. Wiersma, ed., Univ. Chicago Press, Chicago, pp. 39–53.

Autrum, H. (1975) Les yeux et la vision des insects. In *Troite de Zoologie.* VIII. Masson, Paris.

Autrum, H. and D. Burkhardt (1961) Spectral sensitivity of single visual cells. *Nature,* **190,** 639.

Autrum, H. and H. Stumpf (1950) Das Bienenauge als Analysator für polarisiertes. *Licht. Z Naturforsch,* **5b,** 116–122.

Autrum, H. and V. zon Zwehl (1962) Zur spektralen Empfindlichkeit einzelner Sehzellen der Drone *Apis mellifera. Z. Vergl. Physiol,* **46,** 8.

Autrum, H. and V. von Zwehl (1964) Die spektrale Empfindlichkeit einzelner Sehzellen des Bienenauges. *Z. Verg. Physiol.,* **48,** 357.

Axelrod, J. (1974) The pineal gland: A neurochemical transducer. *Science,* **184,** 1341–1348.

Bargmann, W. (1943) *Handbuch der Microskopisehen Anatomie des Menschen,* W. Mollendorf, ed., vol. 6, pp. 309–602, Springer, Berlin.

Barlow, R. B., Jr. (1986) Perception: what quantitative laws govern the acquisition of knowledge from the senses? In *Functions of the Brain* (C. Coen ed.), Clarendon Press, Oxford.

Barlow, R. B., Jr. (1990) What the brain tells the eye. *Sci. Am.,* **262(4),** 90–95.

Bartch, R. G. and M. D. Kamen (1960) Isolation and properties of two soluble proteins in the extracts of the photoanaerobe *Chromatium. J. Biol. Chem.,* **235,** 825.

Baumann, F. A., A. Mauro, R. Milecchia, S. Nightengale, and J. Z. Young (1970) The extraocular receptors of squids *Todarodes* and *Illex. Brain Res.,* **21,** 275–279.

Baylor, E. R. (1967) Air and water vision of the Atlantic flying fish, *Cypselurus necturus*. *Nature*, **214**, 307–309.

Baylor, E. R. and F. E. Smith (1953) A physiological light trap. *Ecology*, **34**, 223–224.

Baylor, E. R. and F. E. Smith (1958) Extra-ocular polarization analysis in the honeybee. *Anat. Rec.*, **132**, 411–412.

Becker, R. S. (1988) The visual process: photophysics and photoisomerization of model visual pigments and the primary reaction. *Photochem. Photobiol.*, **48**, 366–399.

Beebe, W. (1934) Deep-sea fishes of the Bermuda Oceanographic Expeditions, family Idiacanthidae, *Zoologica (New York)*, **16**, 149–241.

Beebe, W. (1935) *Half-Mile Down*, p. 193, Harcourt Brace and Co., New York.

Benjamin, P. R. and J. S. Walker (1972) Two pigments in the brain of freshwater pulmonate snail. *Comp. Biochem. Physiol.*, **41B**, 813–821.

Bennett, M. F. (1979) Extraocular light receptors and circadian rhythms. In *Handbook of Sensory Physiology*, vol. 7/6A, *Vision in Invertebrates, A: Invertebrate Photoreceptors*, H. Autrum, ed., pp. 641–663, Springer-Verlag, Berlin.

Bergelson, L. D. and L. I. Barsukov (1977) Topological asymmetry of phospholipids in membranes. *Science*, **197**, 224.

Bergman, K., P. V. Burke, E. Cedrá-Olmedo, C. N. David, M. Delbrück, K. W. Foster, E. W. Goodell, H. Heisenberg, G. Meissner, M. Zalokar, D. S. Dennison, and W. Shropshire, J. (1969) *Phycomyces. Bacteriol. Rev.*, **33**, 99.

Bertholf, L. M. (1931) The distribution of stimulative efficiency in the ultraviolet spectrum of the honeybee. *J. Agr. Res.*, **43**, 703.

Birge, R. R. (1990) Photophysics and molecular electronic applications of the rhodopsins. *Annu. Rev. Phys. Chem.*, **41**, 683–733.

Blasie, J. K. (1972) Location of photopigment molecules in the cross-section of frog retinal disk membranes. *Biophys. J.*, **12**, 191.

Block, G. D. and M. E. Lickey (1973) Extraocular photoreceptors and oscillators can control the circadian rhythm of behavioral activity in *Aplysia. J. Gen. Physiol.*, **42**, 367–374.

Böll, F. (1876) Zur Anatomie und Physiologie der Retina. *Monatsber. Akad. Wiss., Berlin*, **41**, 738–788.

Bouligand, Y. (1972) Twisted fiber arrangements in biological materials and cholesteric mesophases. *Tissue and Cell*, **4**, 189–217.

Bowmaker, J. K. and A. Knowles (1977) The visual pigments and oil droplets of the chicken retina. *Vision Res.*, **17(7)**, 755–764.

Bowness, J. M. and J. J. Wolken (1959) A light sensitive yellow pigment from the housefly. *J. Gen. Physiol.*, **42**, 779.

Brawerman, G. and J. M. Eisenstadt (1964) DNA from the chloroplast of *Euglena gracilis. Biochem. Biophys. Acta*, **91**, 477.

Bretscher, M. S. (1973) Membrane structures: some general principles. *Science*, **181**, 622.

Brodhun, B. and D. P. Häder (1990) Photoreceptor proteins and pigment in the paraflagelar body of the flagellate *Euglena gracilis. Photochem. Photobiol.*, **52**, 865–871.

Bronowski, J. (1974) *The Ascent of Man*. Little Brown & Company, Boston.

Brown, G. H. and J. J. Wolken (1979) *Liquid Crystals and Biological Structures*. Academic Press, New York.

Brown, P. K. and G. Wald (1964) Visual pigments in single rods and cones of the human retina. *Science*, **144**, 45.

Brown, A. M., P. S. Baur, Jr., and F. H. Tully, Jr. (1975) Phototransduction in *Aplysia* neurons: *Science,* **188,** 157–160.

Bruno, M. S. and D. Kennedy (1962) Spectral sensitivity of photoreceptor neurons in the sixth ganglion of the crayfish. *Comp. Biochem. Physiol.,* **6,** 41–46.

Burkhardt, D. (1962) Spectral sensitivity and other response characteristics of single visual cells in the arthropod eye. In *Biological Receptor Mechanisms,* J. W. L. Beaument, ed., pp. 86–109, Academic Press, New York.

Burkhardt, D. and I. De la Motte (1983) How stalk-eye flies eye stalk-eyed flies: observations and measurements of the eyes of *Cyrtodiopsis whitei* (Diopsidae, Diptera). *J. Comp. Physiol. A.,* **151,** 407–422.

Burkhardt, D. and I. De la Motte (1985) Selective pressures, variability, and sexual dimorphism in stalk-eyed flies (Diopsidae). *Naturwissenschaften,* **72,** 204–206.

Burkhardt, D. and L. Wendler (1960) Ein direkter Bewis für die fähigkeit einzelner Sehzellen des Insektenauges, die Schwingungrsrichtung polarisierten Lichtes zu analysieren. *Z. Vergl. Physiol.,* **43,** 687.

Butenandt, A. and G. Neubert (1955) Über ommochrome V. Xanthommatins ein Augenfarbstoff der Schmeissfliege. *Hoppe-Seyler's Z. Physiol. Chem.,* **301,** 109.

Butenandt, A., V. Schiedt, and E. Bickert (1954) Über Ommachrome. III: Mitteilung Synthese des Xanthommatins. *Justis Liebig's Ann. Chem.,* **588,** 106.

Cajal, R. Ramón y (1918) Observaciones sobre la Estructura de los Ocelos y vias Nerviosas Ocelares le Algunos Insectos. *Trab. Lab. Invest. Biol. Univ. Madrid,* **16,** 109.

Calvin, M. (1962) Path of carbon in photosynthesis. *Science,* **135,** 879.

Calvin, M. (1969) *Chemical Evolution.* Oxford University Press, London.

Calvin, M. (1983) Artificial photosynthesis quantum capture and energy storage. *Photochem. Photobiol.,* **37,** 349–360.

Capaldi, R. A., H. Komap, and D. R. Hunter. (1973) Isolation of a major hydrophobic protein of mitochondrial inner membrane. *Biochem. Biophys. Res. Commun.,* **55,** 655.

Carlson, S. D. (1972) Microspectroscopy of the dioptic apparatus and compound rhabdom of the moth, *Manduca sexta,* eye. *J. Insect Physiol.,* **18,** 1721.

Caveney, S. (1981) Origins of graded index lenses in the superposition eyes of Scarab beetles. *Phil. Trans. Soc. of Lond. B.,* **294,** 589–632.

Cerdá-Olmedo, E. and E. D. Lipson (1987), eds. *Phycomyces.* Cold Spring Harbor Laboratory, Cold Springs Harbor, New York.

Chabre, M. (1975) X-ray diffraction studies of retinal rods. I. Structure of the disc membrane, effect of illumination. *Biochem. Biophys. Acta,* **382,** 332–335.

Chabre, M. (1985) Trigger and amplification mechanismsa in visual transduction. *Annu. Rev. Biophys. Biophysical Chem.,* **14,** 331–360.

Chalazonitis, N. (1964) Light energy conversion in neuronal membranes. *Photochem. Photobiol.,* **3,** 539–559.

Chapell, R. L. and J. E. Dowling (1972) Neural organization of the median ocellus of the dragonfly. *J. Gen. Physiol.,* **60,** 121.

Chapman, D. (1979) Liquid crystals and biological membranes. In *Liquid Crystals,* F. D. Saeva, ed., pp. 305–334, Marcel Dekker, New York.

Chase, R. (1979) Photic sensitivity of the rhinophore in *Aplysia. Can. J. Zool.,* **57,** 698–701.

Clark, R. W. (1989) *Survival of Charles Darwin,* p. 12, Random House, New York.

Clarke, G. L. and E. J. Denton (1962) Light and animal life. In *The Sea*, M. N. Hill, ed., pp. 456–468 Wiley, New York.

Clarkson, E. N. K. and R. Levi-Setti (1975) Tribobite eyes and the optics of DesCartes and Huygens. *Nature*, **254**, 663–667.

Clement, P., E. Wurdak, and J. Amsellem (1983) Behavior and ultrastructure of sensory organs in rotifers. *Hydrobiologia*, **104**, 89–130.

Cohen, D. M. (1959). *Bathylychnops exilis*, a new genus and species of Argentinoid fish from the North Pacific. *Stanford Ichthyology Bulletin*, **7(3)**, 47–52.

Cohen, R. and M. Delbrück (1959) Photoreactions in *Phycomyces*. Growth and tropic responses to the stimulation of narrow test areas. *J. Gen. Physiol.*, **42**, 677.

Cohen, S. S. (1970) Are/were mitochondria and chloroplasts microorganisms? *Am. Sci.*, **58**, 281.

Cohen, S. S. (1973) Mitochondria and chloroplast revisited. *Am. Sci.*, **61**, 437.

Crescitelli, F. (1972) The visual cells and pigments of the vertebrate eye. In *Handbook of Sensory Physiology*, H. J. A. Dartnall, ed., vol. VII, Part 1, pp. 245–363. Springer-Verlag, Berlin.

Crescitelli, F. (1977) The visual pigments of geckos and other vertebrates: an essay in comparative biology. In *Handbook of Sensory Physiology*, vol II, part 5. *The Visual system of vertebrates*, pp. 391–449. Springer-Verlag, Berlin.

Crescitelli, F. and Dartnall, H. J. A. (1953) Human visual purple. *Nature (London)* **172**, 195.

Crick, F. (1988). *What Mad Pursuit*. Harper Collins Publishers, New York.

Croll, N. A. (1966) The phototactic response and spectral sensitivity of *Chromadorina viridis* (Nematoda: Chromadorida) with a note on the nature of paired pigment spots. *Nematology*, **12**, 610–614.

Croll, N. A., A. A. F. Evans, and J. M. Smith (1975) Comparative nematode photoreceptors. *Comp. Biochem. Physiol.*, **51A**, 139–143.

Curry, G. M. and K. V. Thimann (1961) Phototropism: the nature of the photoreceptor in higher and lower plants. In *Progress in Photobiology*, Christensen and Buchmann, eds., pp. 127–134; Elsevier, New York.

Danielli, J. F. and H. Davson (1935) A contribution to the theory of permeability of thin films. *J. Cell. Comp. Physiol.*, **5**, 495.

Dartnall, H. J. A. (1957) *The Visual Pigments*, Wiley, New York.

Dartnall, H. J. A. (1962) Photobiology of the visual process. In *The Eye*, H. Davson, ed., vol. 2, pp. 323–522, Academic Press, New York.

Darwin, C. (1859) *On the Origin of Species by Means of Natural Selection*. Murray, London. (Republished by Modern Library, New York, 1936).

Datta, D. B. (1987) *Membrane Biochemistry*, Floral Publishing Co., Madison, WI.

Daumer, K. (1956) Reizmetrische Untersuchung des Farbensehns der Biene. *Z. Vergl. Physiol.*, **38**, 413.

De Broglie, L. (1955) *Physics and Microphysics*. Pantheon Books, Inc., New York.

Delbrück, M. (1976) Light and life III. *Carlsberg Res. Commun.*, **41**, 299–309.

Delbrück, M. and W. Shropshire, Jr. (1960) Action and transmission spectra of *Phycomyces*. *Plant Physiol.*, **35**, 194.

Del Priore, L. V., A. Lewis, S. Tan, W. W. Carley, and W. W. Webb (1987) Flourescence light microscopy of F-actin in retinal rods and glial cells. *Investigative Ophthalmol. Vis. Sci.*, **28**, 633–639.

Denton, E. J. (1990) Light and vision at depths greater than 200 meters. In *Light and Life in the Sea,* P. J. Herring, A. K. Campbell, M. Whilfred, and L. Maddoch, eds., Cambridge University Press, Cambridge.

de Saussure, N. T. (1804) *Recherches chimiques sur la végétation,* Lyon, France.

Descartes, R. (1637) *Descartes: Discourse on Method, Optics, Geometry and Meterology,* P. J. Olscamp, trans., A. Leyde, De l'imprimerie de I Maire. (Republished by Bobbs-Merrill, New York, 1965). (See Smith, 1987.)

Detwiler, S. R. (1943) *Vertebrate Photoreceptors,* Macmillan, New York.

Diehn, B. (1973) Phototaxis and sensory transduction in *Euglena. Science,* **181,** 1009.

Dilly, P. N. (1964) Studies on the receptors in the cerebral vesicle of the ascidian tadpole. II. The Ocellus *Q. J. Microsc. Sci.* [*N.S.*], **105,** 13.

Dilly, P. N. (1969) Studies on the receptors in *Ciona intestinalis.* III. A second type of photoreceptor in the tadpole larva of *Ciona intestinalis.* Z. Zellforsch. Mikrosk. Anat., **96,** 63.

Dilly, P. N. and J. J. Wolken (1973) Studies on the photoreceptors in *Ciona intestinalis.* IV. The ocellus of the adult. *Micron,* **4,** 11–29.

Dobell, C. (1958) *Antony van Leeuwenhoek and His "Little Animals",* Russell & Russell, Inc., New York.

Dodt, E. and M. Jacobson (1963) Photosensitivity of a localized region of the frog diencephalon. *J. Neurophysiol.,* **26,** 752–758.

Donner, K. O. (1953) The spectral sensitivity of the pigeon's retinal elements. *J. Physiol. (London),* **122,** 524.

Dougherty (1993) Photodynamic therapy. *Photochem. and Photobiol.,* **58,** 895–900.

Døving, K. B. and W. H. Miller (1969) Function of insect compound eyes containing crystalline tracts. *J. Gen. Physiol.,* **54,** 250.

Dowling, J. E. (1987) *The Retina.* The Belknap Press of Harvard University, Cambridge, MA.

Dowling, J. E. and H. Ripps (1970) Visual adaptation in the retina of the skate. *J. Gen. Physiol.,* **56,** 491–520.

Duke-Elder, S. (1958) *System of Ophthalmology: The Eye in Evolution,* vol. 1, p. 102, Mosby, St. Louis.

Dulbecco, R. (1949) Reactivation of ultraviolet inactivated bacteriophage by visible light. *Nature,* **163,** 949–950.

Dumortier, B. (1972) Photoreception in the circadian rhythm of stridulatory activity in *Ephippiger. J. Comp. Physiol.,* **77** 81–112.

Duysens, L. N. M. (1964) Photosynthesis. In J. A. V. Butler and H. E. Huxley, eds. vol. 14 Pergamon Press, London. *Progress in Biophysic,* **14,** 2–104.

Eakin, R. M. (1963) Lines of evolution of photoreceptors. In *General Physiology of Cell Specialization,* D. Mazia and A. Tyler, eds., pp. 393–425, McGraw-Hill, New York.

Eakin, R. M. (1965) Evolution of photoreceptors. *Cold Spring Harbor Symp. Quant. Biol.,* **30,** 363.

Eakin, R. M. (1973) *The Third Eye.* University of California Press, Berkeley.

Eakin, R. M. and A. Kuda (1971) Ultrastructure of sensory receptors in ascidian tadpoles. *Z. Zellforsch. Mikrosk. Anat.,* **112,** 287.

Eguchi, E. and T. H. Waterman (1966) Fine structure patterns in crustacean rhabdoms. In *Functional Organization of the Compound Eye,* C. G. Bernhard, ed., pp. 105–124, Pergamon Press, Oxford.

Ehrenkranz, J. R. L. (1983) A gland for all seasons. *Natural History,* **6,** 19–23.

Eichenbaum, D. M. and T. H. Goldsmith (1968) Properties of intact photoreceptor cells lacking synapses. *J. Exp. Zool.*, **169**, 15.

Einstein, A. (1905) On a heuristic viewpoint concerning the production and transformation of light. *Ann. Phys. (Leipzig)*, **17**, 132.

Eisberg, R. M. and R. Resnick (1985) *Quantum Physics,* Wiley, New York.

Eisenburg, M. and S. McLaughlin (1976) Lipid bilayers as models of biological membranes. *BioScience*, **26**, 436.

Eldred, W. D. and J. Nolte (1978) Pineal photoreceptors: evidence for a vertebrate visual pigment with two physiologically active states. *Vis. Res.*, **18**, 29–32.

Emerson, R. (1956) Effect of temperature on long-wave limit of photosynthesis. *Science,* **123**, 637.

Emerson, R. and C. M. Lewis (1943) The dependence of the quantum yield of *Chlorella* photosynthesis on wavelength of light. *Am. J. Bot.*, **30**, 165.

Enoch, J. M., and F. L. Tobey, Jr. (1981) *Wave Guide Properties of Retinal Receptors in Vertebrate Photoreceptor Optics.* pp. 169–218. Springer-Verlag, New York.

Engbretson, G. A. and C. M. Lent (1976) Parietal eye of the lizard: neuronal photoresponses and feedback from the pineal gland. *Proc. Natl. Acad. Sci. U.S.A.*, **73**, 654–657.

Englemann, T. W. (1882) Uber Licht- und Farbenperception neiderster Organismen. *Pflügers Arch. Gesamte Physiol., Menschen Tiere*, **29**, 387.

Engström, K. (1963) Cone types and cone arrangements in teleost retinae. *Acta. Zool.*, **44**, 179–243.

Exner, S. (1876) Uber das Sehen von Bewegungen und die Theorie des zusammengesetzten Auges. *Sitz. Ber. Kaiserl. Akad. Wiss. Math. Nat. Wiss.*, **72**, 156–191.

Exner, S. (1891). Die Physiologie du Facetteir ten Augen Non Krebsen Und Insecten. Leipzig, Germany and Franz Deuticke Vienna, Austria. Translated as *The Physiology of Compound Eyes in Insects and Crustacea,* in German, R. G. Hardie, trans., Springer-Verlag, Berlin, New York.

Fauré-Frémiet (1958) The origin of the metazoa and the stigma of phytoflagellates. *Q. J. Microsc. Sci.*, **99**, 123.

Fernald, R. D. (1990) Optical system of fishes. In *The Visual Systems of Fish*, R. Douglas and M. Djamgoz, eds. Chapman and Hall, New York.

Fernald, R. D. and S. E. Wright (1983) Maintenance of optical quality during crystalline lens growth. *Nature,* **301**, 618–620.

Fernandez-Morán, H. (1956) Fine structure of the insect retinula as revealed by electron microscopy. *Nature (London)*, **177**, 742.

Fernandez-Morán, H. (1958) Fine structure of the light receptors in the eyes of insects. *Exp. Cell Res. Suppl.*, **5**, 586.

Fingerman, M. (1952) The role of the eye pigments of *Drosophila melanogaster* in photic orientation. *J. Exp. Zool.*, **120**, 131.

Fingerman, M. and F. A. Brown (1952) A "Purkinje shift" in insect vision. *Science,* **116**, 171.

Fingerman, M. and F. A. Brown (1953) Color discrimination and physiological duplicity of *Drosophila* vision. *Physiol. Zool.*, **26**, 59.

Fischer, H. and A. Stern (1940) *Die Chemie de Pyrrols,* vol. 2, Hälfte 2, Akad. Verlagsges., Leipzig.

Fletcher, A., T. Murphy, and A. Young (1954) Solutions of two optical problems. *Proc. R. Soc. London,* **223**, 216–250.

Forrest, H. S. and H. K. Mitchell (1954) Pterdines from *Drosophila*. I. Isolation of a yellow pigment. *J. Am. Chem. Soc.*, **76**, 5656.

Foster, K. W. and R. D. Smyth (1980) Light antennas in phototactic algae. *Microbiol. Rev.*, **44**, 572–630.

Foster, K. W., J. Saranak, N. Patal, G. Zarilli, M. Okabe, T. Kline, and K. Nakanishi (1984) A rhodopsin is the functional photoreceptor for phototaxis in the unicellular eukaryote *Chlamydomonas*. *Nature*, **311**, 756.

Fox, D. L. (1953) *Animal Biochromes and Structural Colors*, Cambridge Univ. Press, London (2nd ed., Univ. California Press., Los Angeles, 1976).

Galland, P. and E. D. Lipson (1984) Photophysiology of *Phycomyces blakesleeanus*. *Photochem. Photobiol.*, **40**, 795–800.

Galland, P. and E. D. Lipson (1985) Action spectra for phototropic balance in *Phycomyces blakesleeanus:* dependence on reference wavelength and intensity range. *Photochem. Photobiol.*, **41**, 323–329.

Geethabali, X. and K. P. Rao (1973) A metasomatic neural photoreceptor in the scorpion. *J. Exp. Biol.*, **58**, 189–196.

Gilbert, P. W. (1963) The visual apparatus of sharks. In *Sharks and Survival*, P. W. Gilbert, ed., pp. 283–326, Heath, Lexington, MA.

Glover, J., T. W. Goodwin, and R. A. Morton (1948) Conversion in vivo of vitamin A aldehyde (retinene$_1$) to vitamin A$_1$. *Biochem. J.*, **43**, 109.

Goldsmith, T. H. (1958) On the visual system of the bee (*Apis mellifera*). *Ann. N.Y. Acad. Sci.* **74**, 223.

Goldsmith, T. H. (1960) The nature of the retinal action potential and the spectral sensitivities of ultraviolet and green receptor systems of the compound eye of the worker honeybee. *J. Gen. Physiol.*, **43**, 775.

Goldsmith, T. H. (1962) Fine structure of the retinulae in the compound eye of the honeybee. *J. Cell Biol.*, **14**, 489.

Goldsmith, T. H. (1975) The polarization sensitivity–dichroic absorption paradox in arthropod photoreceptors. In *Photoreceptor Optics*, A. W. Snyder and R. Menzel eds. pp. 63–114, Springer, Berlin.

Goldsmith, T. H. (1986) Interpreting trans-retinal recordings of spectral sensitivity. *J. Comp. Physiol.*, **159**, 481–487.

Goldsmith, T. H. (1990) Optimization, constraint, and history in the evolution of the eyes. *Q. Rev. Biol.*, **65**, 281–322.

Goldsmith, T. H. and D. E. Philpott (1957) The microstructure of the compound eyes of insects. *J. Biophys. Biochem. Cytol.*, **3**, 429.

Goodwin, D. W. and G. Britton (1988) *Plant Pigments*, T. Goodwin, ed., Academic Press, London.

Gotow, R. (1975) Morphology and function of photoexcitable neurons in the central ganglia of *Onchidium verruculatum*. *J. Comp. Physiol.*, **99**, 139–157.

Granick, S. (1948) Magnesium protoporphyrin as a precursor of chlorophyll in *Chlorella*. *J. Biol. Chem.*, **175**, 333.

Granick, S. (1950) Magnesium vinyl pheoporphyrin a$_5$, another intermediate in the biological synthesis of chlorophyll. *J. Biol. Chem.*, **183**, 713.

Granick, S. (1958) Porphyrin biosynthesis in erythrocytes. I. Formation of D-aminolevulinic acid in erythrocytes. *J. Biol. Chem.*, **232**, 1101.

Greef, R. (1877) *Monograph Alciopidae*, 39, No. 2, Nova Acta Kais, Leopold.

Gregory, R. L. (1966) *Eye and Brain*, McGraw-Hill, New York.

Gregory, R. L. (1967) Origin of eyes and brains. *Nature*, **213**, 369.

Grenacher, H. (1879) *Untersuchungen über das Sehorgan der Arthropoden, insbesondere der Spinner, Insekten and Crustaceen,* p. 195 Vanderhoeck and Ruprecht, Gottingen, Germany.

Grenacher, H. (1886) Abhandlungen zur vergleichen Anatomie des Auges I. Die Retina der Cephalopoden. *Abh Naturforsch. Ges Halle,* **16**, 207.

Grossbach, U. (1957) Zur papierchromatographischen Untersuchung von Lepidopteren-Augen. *Z. Naturforsch,* **12b**, 462.

Gruber, S. H., D. I. Hamasaki, and C. D. B. Bridges (1963) Cones in the retina of the lemon shark (*Negaprion brevirotris*). *Vis. Res.,* **3**, 397.

Gust, D. and T. A. Moore (1991) Mimicking photosynthetic electron and energy transfer. In *Advances in Photochemistry,* D. Volman, G. Hammond, and D. Nechers, eds., vol. 16, Wiley, New York.

Gwilliam, G. F. (1969) Electrical responses to photo-stimulation in the eyes and nervous system of nerid polychaetes. *Biol. Bull.,* **16**, 385–397.

Haidinger, W. (1844) Über das direkte Erkennen des polarisierten Lichts und Lage der Polarisationsebenne *Ann. Physik.* **63**, 29.

Haldane, J. B. S. (1966) *The Causes of Evolution,* Cornell Univ. Press, Ithaca, N.Y.

Hamasaki, D. I. and P. Streck (1971) Properties of the epiphysis cerebri of the small spotted dogfish shark *Scyliohinus caniculus* L. *Vis. Res.,* **11**, 189–198.

Hamdorf, H. (1979) The physiology of invertebrate visual pigments. In *Handbook of Sensory Physiology,* Vol. VII 16 A, H. Autrum, ed., Springer, Berlin.

Hanaoka, T. and K. Fujimoto (1957) Absorption spectrum of a single cone in the carp retina. *Jpn. J. Physiol.,* **7**, 276.

Hara, T. and R. Hara (1980) Retinochrome and rhodopsin in the extraocular photoreceptor of the squid *Todarodes*. *J. Gen. Physiol.,* **75**, 1–19.

Harth, M. S. and M. B. Heaton (1973) Nonvisual photo responsiveness in newly hatched pigeons (*Columbia livia*). *Science,* **180**, 753–755.

Hartline, H. K. ed. (1974) *F. Ratliff and Associates, Collected Papers,* Rockefeller Univ., New York.

Hartwig, H. G. and C. Baumann (1974) Evidence for photosensitive pigments in the pineal complex of the frog. *Vis. Res.,* **14**, 597–598.

Harvey, E. N. (1960) Bioluminescence. *Comp. Biochem.,* **2**, 545–591.

Hasegowa, K. and I. Shimizu (1988) Occurrence of retinal and 3-hydroxyretinal in a possible photoreceptor of the silkworm brain involved in photoperiodism. *Experientia,* **44**, 74–76.

Hawkins, E. G. E. and R. F. Hunter (1944) Vitamin A aldehyde. *J. Chem. Soc. (London),* 411.

Hawryshyn, C. W. (1992) Polarization vision in fish. *Am. Sci.,* **80**, 164–175.

Helbig, A. J. and W. Wiltschko (1989) The skylight polarization of patterns at dusk affect the orientation behavior of blackcaps, *Sylvia atricappila, Naturwissenschaften,* **76**, 227–229.

Hendricks, S. B. (1968) How light interacts with living matter. *Sci. Am.,* **219**, 174.

Hendricks, S. B. and H. W. Siegleman (1967) Phytochrome and photoperiodism in plants. *Comp. Biochem.,* **27**, 211–235.

Hering, E. (1885) *Ueber Individuelle Verachiedenbeiten des Farbensinnes,* Lotos, Prague. And in *Outlines of a Theory of the Light Sense,* trans. L. M. Hurvich and D. Jameson, Harvard Univ. Press, Cambridge, MA, 1965.

Hermans, C. O. and R. M. Eakin (1974) Fine structure of eyes of an alciopid polychaete, *Vanadis tangensis* (Annelida). *Z. Morphol. Tierre,* **79,** 245–267.

Hertz, H. R. (1894–95) *Gesammelte Werke.* J. A. Barth, Leipzig.

Hertz, M. (1939) New experiments on color vision in bees. *J. Exp. Biol.,* **16,** 1.

Hesse, R. (1899) Untersuchungen über die Organe der Lichtenpfindung bie niederen Thieren. V. Die Augen den polychaten Anneliden, A. *Wiss. Zool.,* **65,** 446–516.

Hill, R. and F. Bendall (1960) Function of the two cytochrome components in chloroplasts: a working hypothesis. *Nature,* **186,** 136.

Hillman, P., S., Hockstein, and B. Minke (1983) Transduction in invertebrate photoreceptors: role of pigment bistability. *Physiol. Rev.,* **63,** 668–672.

Hisano, N., H. Tateda, and M. Kubara (1972a) Photosensitive neurons in the marine plumonate mollusk *Onchidium verruculatum. J. Exp. Biol.,* **57,** 651–660.

Hisano, N., H. Tateda, and M. Kubara (1972b) An electrophysiological study of the photoexcitable neurons of *Onchidium verruculatum. J. Exp. Biol.,* **57,** 661–671.

Hisano, N., D. P. Cardinalli, J. M. Rosner, C. A. Nagle, and J. H. Tremezzani (1972c) Pineal role in the duck extraretinal photoreception. *Endocrinology* **91,** 1318–1322.

Hooke, R. (1665) *Micrographia.* Reprinted in *Early Science in Oxford,* R. T. Gunther, ed., 1938, Oxford Univ. Press, Oxford.

Hopkins, S. F. G. (1889) Note on a yellow pigment in butterflies. *Nature,* **40,** 335.

Horridge, G. A. (1966) Arthrodopoda: receptor for sight and optic lobes. In *Structure and Function in the Nervous System of Invertebrates* T. H. Bullock and G. H. Horridge, eds., Freeman, San Francisco.

Horridge, G. A. (1968) Pigment movement and the crystalline threads of the firefly eye. *Nature,* **218,** 778.

Horridge, G. A. (1969) The eye of the firefly *Photuris. Proc. R. Soc. Ser.,* **B171,** 445.

Horridge, G. A. (1975) *The Compound Eye and the Vision of Insects,* Oxford Univ. Press, New York.

Horridge, G. A. (1986) A theory of insect vision. *Proc. R. Soc. Lond., B.,* **229,** 13–27.

Hoyle, Fred (1967) *The Black Cloud.* Penguin Books, England.

Hubbard, R. (1954) The molecular weight of rhodopsin and the nature of the rhodopsin-digitonin complex. *J. Gen. Physiol.,* **37,** 381.

Hubbard, R., P. K. Brown, and D. Bownds (1971) Methodology of vitamin A and visual pigments. In *Methods in Enzymology,* D. B. McCormick and L. D. Wright, eds., vol. 18C, pp. 615–653, Academic Press, New York.

Hubel, D. and T. Weisel (1979) Brain mechanisms of vision. *Sci. Am.* **241,** 150–162.

Hunter, R. F. and N. E. Williams (1945) Chemical conversion of β-carotene into vitamin A. *J. Chem. Soc. (London),* **2,** 554–555.

Hurvich, Leo M. (1981) *Color vision,* Sinauer Associates, Inc., Sunderland, MA.

Huygens, C. (1678) *Traite de la lumière. Ou sont expliquées les causes de ce qui luy arrive dans la réfléxion, & dans la réfraction. Et particulierement dans l' étrange réfraction du cristal d'Islande,* A. Leide, chêz Pierre van der Aa, marchand libraire. MDCXC.

Ingen-Housz, J. (1779) *Experiments upon Vegetables, Discovering their Great Power of Purifying the Common Air in Sunshine and Injuring it in the Shade and at Night.* Elmsly & Payne, London.

Ingen-Housz, J. (1798) *Ernährung der Pflanzen und Fruchtbarkeit des Bodens,* Leipzig.

Jackson, R. R. (1985) A web-building jumping spider. *Sci. Am.,* **253,** 102–115.

Jacobs, G. H. (1993) Distribution and nature of color vision among the mammals. *Biol. Rev.,* **68,** 413–471.

Jacobs, G. H., J. Neitz, and J. F. Deegan, II (1991) Retinal receptors in rodents maximally sensitive to ultraviolet light. *Nature,* **353,** 655–656.

James, T. W., F. Crescitelli, E. R. Loew, and W. N. McFarland (1992) The eyespot of *Euglena gracilis:* A Microspectrophotometric Study. *Vis. Res.,* **32,** 1583–1591.

Jennings, H. S. (1906). *Behavior of the Lower Organisms.* Columbia Univ. Press, New York.

Jenkins, F. A. and H. E. White (1976) *Fundamentals of Optics,* McGraw-Hill Publishing Co., New York.

Jerlov, N. G. (1976) *Marine Optics,* Elsevier Publishing Co. Amsterdam, The Netherlands.

Jordan, P. M. (1991) *Biosynthesis of Tetrapyrroles,* Elsevier Science Publishers, New York.

Jorschke, H. (1914) Die Facettenaugen der orthopteren und termiten. *Z. Wiss. Zool.,* **111,** 153.

Journeaux, R. and R. Viovy (1978) Orientation of chlorophyll in liquid crystals. *Photochem. Photobiol.,* **28,** 243.

Justis, C. S. and D. H. Taylor (1976) Extraocular photoreception and compass orientation in larval bullfrogs *Rana catesbeiana. Copeia,* **1,** 98–125.

Kamen, M. D. (1956) Hematin compounds in metabolism of photosynthetic tissues. In *Enzymes: Units of Biological Structure and Function,* O. H. Gaebler, ed., pp. 483–504, Academic Press, New York.

Kamen, M. D. (1960) Hematin compounds in photosynthesis. In *Comparative Biochemistry of Photoreactive Systems,* M. B. Allen, ed., pp. 323–327, Academic Press, New York.

Karrer, P. and E. Jucker (1950) *Carotenoids,* Elsevier, New York.

Kästner, A. (1950) Reaktion der Hupfspinnen (Salticidae) auf unbewegte farblose und farbige gesichtsreige. *Zool. Beitr. (Berlin), [N.S.],* **1,** 12.

Ke, B. and L. Vernon (1971) Living systems in photochromism. In *Photochromism,* G. H. Brown, ed. p. 687, Wiley, New York.

Keck, D. B. and R. E. Love. (1980) *Fiber Optics for Communications.* In *Applied Optics and Optical Engineering,* R. Kingslake and B. J. Thompson, eds. vol. 6, p. 439, Academic Press, New York.

Keeble, F. (1910) *Plant-Animals: A Study in Symbiosis.* Cambridge Univ. Press, Cambridge.

Kelly, D. (1965) Ultrastructure and development of amphibian pineal organs. *Prog. Brain Res.,* **10,** 270.

Kelly, D. (1971) Developmental aspects of amphibian pineal systems, *Pineal Gland. Ciba Found. Symp., 1970,* G. E. Wolstenholme and J. Knight, eds. pp. 53–77, Churchill Livingstone, Edinburgh.

Kelner, A. (1949) Photoreactivation of ultraviolet-radiated *Escherichia coli,* with special reference to the dose-reduction principle and to ultra-violet mutation. *J. Bacteriol.,* **58,** 511–522.

Kelner, A. (1949a) Effect of visible light on the recovery of *Streptomyces grisens conidia* from ultraviolet irradiation injury. *Proc. Natl. Acad. Sci. U.S.A.,* **35,** 73–79.

Kelner, A. (1949b) Photoreactivation of ultraviolet-irradiated *Escherichia coli,* with special reference to the dose-reduction principle and to ultraviolet-induced mutation. *J. Bacteriol.,* **58,** 511–522.

Kennedy, D. and E. R. Baylor (1961) The spectral sensitivity of crayfish and lobster vision. *J. Gen. Physiol.,* **44,** 1089.

Kiewisch, S. and L. Fukshansky (1991) Detection of pterins in *Phycomyces* sporangiophores. *Photochem. Photobiol.*, **53**, 407–409.

Kingslake, R. (1983) *Optical System Design.* Academic Press, New York.

Kofoid, C. A. and C. Swezy (1921) *Free Living Unarmored Dinoflagellate.* Memoire. Univ. California, **5**, Berkeley.

Können, G. P. (1985) *Polarized Light in Nature,* trans. G. A. Bearling, Cambridge Univ. Press, Cambridge.

Korn, E. D. (1964) The fatty acids of *Euglena gracilis. J. Lipid Res.*, **5**, 352.

Korn, E. D. (1966) Structure of biological membranes. *Science*, **153**, 1491–1498.

Kraughs, J. M., L. A. Sordhal, and A. M. Brown (1977) Isolation of pigment granules involved in extra-retinal photoreception in *Aplysia californica* neurons. *Biochem. Biophys. Acta*, **471**, 25–31.

Krause, W. (1863) Uber die Endigung der Muskelnerven. *Z. Ration. Med.*, **20**, 1.

Kreimer, G. (1994) Cell biology of phototaxis in flagellate algae. In *International Review of Cytology*, Vol. 148, K. W. Jeon and J. Jarvik, eds., pp. 229–310, Academic, San Diego, CA.

Kreithen, M. L. and T. Eisner (1978) Ultraviolet light detection in the homing pigeon. *Nature*, **272**, 347–348.

Krinsky, N. I. (1971) Function of carotenoids. In *Carotenoids*, O. Isler, ed., pp. 669–716, Birkhauser Verlag, Basel.

Krohn, A. (1842) Nachtragliche Beobachtungen uber den Bau des Auges der Cephalopoden. *Verh. Kaiserlich Leopoldenisch-Carolinisch Deutsch Acad. Naturforsch*, **XIX(2)**, 41–50.

Kühn, H. (1927) Über den farbensinn der bienen. *Z. Vergl. Physiol.*, **5**, 762.

Kühn, A. (1968) On possible ways of assembling simple organized systems of molecules. In *Structural Chemistry and Molecular Biology*, A. Rich and N. Davidson, eds., pp. 566–572, Freeman, San Francisco.

Kühne, W. (1878) *On the Photochemistry of the Retina and on Visual Purple*, M. Foster, ed., Macmillan, New York.

Kushmer, D. J. (1985) The Halobacteriaceae. In *The Bacteria VIII. Archaebacteria*, J. R. Sokatch, L. N. Ornston, C. R. Woese, and R. S. Wolfe, eds. Academic Press, Orlando, Florida.

Kuwabara, M. (1957) Bildung des bedingten reflexes von paulous typus bei der honigbiene *Apis mellifera. J. Fac. Sci., Hokaido Univ.*, Ser. 6, **13**, 458.

Land, M. F. (1980) Eye movements and the mechanism of vertical steering in euphausiid Crustacae. *J. Comp. Physiol. (A)*, **137**, 255–265.

Land, M. F. (1980) Optics and vision in invertebrates. In *Handbook of Sensory Physiology*, VII/6B. H. Autrum, ed., pp. 471–592, Springer-Verlag, Berlin, New York.

Land, M. F. (1981) Optics and vision in invertebrates. In *Handbook of Sensory Physiology*, H. Autrum, ed., Vol VII/6B, pp. 471–592, Springer-Verlag, Berlin.

Land, M. F. (1985) Morphology and optics of spider eyes. In *Neurobiology of Arachnids*, F. G., Barth, ed., pp. 53–78, Springer-Verlag, Berlin.

Land, M. F. (1989) The eyes of hybrid amphipods: relations of optical structure to depth. *J. Comp. Physiol. (A)*, **164**, 752–762.

Land, M. F. (1990) Direct observation of receptors and images in simple and compound eyes. *Vis. Res.*, **30**, 1721–1734.

Land, M. F. (1992) The evolution of eyes. *Annu. Rev. Neurosci.*, **15**, 1–29.

Langer, H. (1967) Über die pigmentgranula im facettenaugen von *Callifora erythro-cephala. Z. Vergl. Physiol.,* **55,** 354.

Langer, H., G. Schmeinck, and F. H. Anton-Erkleben (1986) Indification and localization of visual pigments in the retina of the moth *Antheraea polyphemus* (Insecta, Satur-niidae). *Cell Tissue Res.* **245,** 81–89.

Langer, H. and B. Thorell (1966) Microspectrophotometry of single rhabdomeres in the insect eye. *Exp. Cell Res.,* **44,** 673.

Lanyi, J. K. (1992) Proton transfer and energy coupling in bacteriorhodopsin photocycle. *J. Bioeng. Biomemb.,* **24,** 169–179.

Larimer, J. L., D. L., Trevino, and F. A. Ashby (1966) A comparison of spectral sensi-tivities of caudal photoreceptors of epigeal and cavenicolous crayfish. *Comp. Bio-chem. Physiol.,* **19,** 409–415.

Lassansky, A. (1967) Cell junctions in ommatidium of *Limulus. J. Cell Biol.,* **33,** 365.

Lavoisier, A. L. (1774) *Opuscules Physique et Chimiques.* Durand, Paris.

Layll, H. (1957) Cone arrangements in teleost retinae. *Q. J. Micro. Sci.,* **98,** 189–202.

Leduc, S. (1911) *The Mechanism of Life.* Rebman, New York.

Lentz, T. L. (1968) *Primitive Nervous Systems.* Yale Univ. Press, New Haven, CT.

Lerner, A. B., J. D. Case, Y. Takahashi, T. H. Lee, and W. Mori (1958) Isolation of melatonin, the pineal gland factor that lightens melanocytes. *J. Am. Chem. Soc.,* **80,** 2587.

Levi-Setti, R. (1975) *Trilobites: A Photographic Atlas,* Univ. Chicago Press, Chicago.

Lewin, R. A., ed. (1962) *Physiology and Biochemistry of Algae.* Academic Press, New York.

Lewis, A. and L. V. Del Priore (1987) The biophysics of visual photoreceptors. *Physics Today,* **41,** 38–46.

Liaaen-Jenson, S. and A. G. Andrews (1972) Microbial carotenoids. *Annu. Rev. Micro-biol.,* **26,** 225.

Lickey, M. E., G. D. Block, D. J. Hudson, and J. T. Smith (1976) Circadian oscillators and photoreceptors in the gastropod, *Aplysia. Photochem. Photobiol.,* **23,** 253–257.

Lickey, M. E. and S. Zack (1973) Extraocular photoreceptors can entrain the circadian rhythm in the abdominal ganglion of *Aplysia. Photochem. Photobiol.,* **86,** 361–366.

Liebman, P. A. and G. Entine (1964) Sensitive low-light level of microspectrophotometric detection of photosensitive pigment of retinal cones. *J. Opt. Soc. Am.,* **54,** 1451.

Liebman, P. A. and A. M. Granda (1971) Microspecrophotometric measurement of visual pigments in two species of turtle, *Pseudemys scripta* and *Chelomia mydas. Vis. Res.,* **11,** 105–114.

Linzen, B. (1959) Über die verbreitung der ommochrome, der dunkeln augenpigments der arthropoden. *Zool. Anz.,* Suppl. **22,** 154.

Lipetz, L. E. (1984) Pigment types, densities and concentrations in cone oil droplets of *Emydoidea blandingii. Vis. Res.,* **24,** 605.

Lipson, E. D. (1983) Sensory information processing at the cellular level: the light response systems of *Phycomyces.* In *Radiation and Cellular Response,* G. P. Scott and H. W. Wahner, eds., pp. 138–152, Iowa State Univ. Press, Ames.

Lorenz, L. (1867) *Ouvres Scientifiques* (1904) Vol. II. Publices aux Frais de la Fondation Carlesburg. Copenhagen, Denmark.

Lubbock, J. (1882) *Ants, Bees and Wasps,* Appleton, New York.

Lumière, A. and L. Lumière (1894). Note sure la photographie des couleurs. Académie des

Sciences de Lyon, en Principaux Travaux Auguste Lumière, Leon Sezanne, Lyon, 1928.

Luneberg, R. K. (1944) *Mathematical Theory of Optics,* pp. 208–213. Brown Univ. Graduate School, Providence, RI. (Republished by Univ. California Press, Berkeley, 1964).

Lythgoe, J. N. (1979) *The Ecology of Vision.* Clarendon Press, Oxford.

Lythgoe, J. N. and J. C. Partridge (1989) Visual pigments and the acquisition of visual information. *J. Exp. Biol.,* **146,** 1–21.

Manning, J. E. and O. C. Richards (1972) Synthesis and turnover of *Euglena gracilis* nuclear and chloroplast deoxyribonucleic acid. *Biochemistry,* **11,** 2036.

Marak, G. E. and J. J. Wolken (1965) An action spectrum of the fire ant: *Solenopsis saevissima. Nature,* **205,** 1328.

Margulis, L. (1970) *Origin of Eucaryotic Cells.* Yale Univ. Press, New Haven, CT.

Margulis, L. (1982) *Early Life.* Science Books International, Boston.

Marks, P. S. (1976) Nervous control of light responses in the sea anemone, *Calamactis praelongus. J. Exp. Biol.,* **65,** 85–96.

Marks, W. B. (1963) Different spectra of the visual pigments in single goldfish cones. Thesis, Johns Hopkins Univ., Baltimore.

Marks, W. B., W. H. Dobele, and J. B. MacNichol (1964) Visual pigments of single primate cones. *Science,* **143,** 1181.

Marr, D. (1982) *Vision: Processing of Visual Information,* Freeman, San Francisco.

Martin, G. R. and W. R. A. Muntz (1978) Retinal oil droplets and vision in the pigeon (*Columba livia*). In *Neural Mechanisms of Behavior in the Pigeon.* pp. 620–621, Plenum, New York.

Masland, R. H. (1986) The functional architecture of the retina. *Sci. Am.* **6,** 103.

Mast, S. O. (1911) *Light and the Behavior of Organisms,* Wiley, New York.

Mathis, P. and A. W. Rutherford (1987) The primary reactions of photosynthesis I and II of algae and higher plants. In *Photosynthesis,* J. Amesz, ed., Elsevier Science Publishers, New York.

Matthiessen, L. (1886) Ueber den physikalisch-optiken Blau des Auges der Cetacean und der Fische. *Phlügers Arch. ges. Physiol.,* **36,** 521–528.

Mauro, A. (1977) Extraocular photoreceptors in cephalopods. *Symp. Zool. Soc., London,* **38,** 287–308.

Mauro, A. and F. Baumann (1968) Nervous control of light responses of photoreceptors in the epistellar body of *Eledone moschata. Nature,* **220,** 1332–1334.

Mauro, A. and A. Sten-Knudsen (1972) Light-evoked impulses from extraocular photoreceptors in the squid *Todarodes. Nature,* **237,** 342–343.

Maxwell, J. C. (1853) Some solutions to problems. In *Scientific Papers of J. Clerk Maxwell,* W. D. Niven, ed., vol. 1, Dover Books, New York, 1952.

Maxwell, J. C. (1861) On the theory of compound colors and the relations of the colours of the spectrum. *R. Soc. London Phil. Trans.,* **150,** 70.

Maxwell, J. C. (1864) A dynamic theory of the electromagnetic field. In *Scientific Papers of J. Clerk Maxwell,* W. D. Niven, ed., vol. 1, Dover Books, New York, 1952.

Maxwell, J. C. (1890) The theory of compound colours and the relations of those colours to the spectrum. In *The Scientific Papers of J. Clerk Maxwell,* W. D. Niven, ed., vol. I, Dover Books, New York, 1952.

Mayer, R. (1845) *Die Organische Bewegung in ihrem Zusammenhang mit dem Stoffwechsel.* C. Dreshsler, Hielbronn.

Mayr, E. (1982) *The Growth of Biological Thought*. Harvard Univ. Press, Cambridge, MA.

McCrone, E. J. and P. G. Sokolove (1979) Brain-gonad axis and photoperiodically stimulated sexual maturation in the slug, *Limax maximus*. *J. Comp. Physiol*. **133**, 117–123.

McCulloch, W. S. (1965) *Embodiments of Mind*. pp. 230–255, MIT Press, Cambridge, MA.

McMillan, J. P., J. A. Elliot, and M. Menaker (1975a) On the role of eyes and brain photoreceptors on the sparrow: Schoff's rule. *J. Comp. Physiol.*, **102**, 257–262.

McMillan, J. P., J. A. Elliot, and M. Menaker (1975b) On the role of eyes and brain photoreceptors in the sparrow: arhythmicity in constant light. *J. Comp. Physiol.*, **102**, 263–268.

Meissner, G. and M. Delbrück (1968) Carotenes and retinal in *Phycomyces* mutants. *Plant Physiol.*, **43**, 1279.

Menaker, M., ed. (1976) Extra-retinal photoreception. *Photochem. Photobiol.*, **23(4)**, 231–306.

Menaker, M. (1977) Extra-retinal photoreceptors. In *The Science of Photobiology*, K. C. Smith, ed., pp. 227–240, Plenum, New York.

Menaker, M. and H. Underwood (1976) Extraretinal photoreception in birds. *Photochem. Photobiol.*, **23**, 299–306.

Menner, E. (1938) Die Bedeutung des Pecten im Auge des Vogels für die Wahrnehmung von Bewegungen, nebst Bemerkungen über seine Ontogenie und Histologie. *Zool. Jahrb. Abt. Allg. Zool. u. Physiol. Tiere*, **58(4)**, 481–538.

Menzel, R., D. F. Ventura, H. Hertel, J. de Souza, and U. Greggers (1986) Spectral sensitivity of photoreceptors in insect compound eyes: comparison of species and methods. *J. Comp. Physiol. A*, **158**, 165–177.

Meyer, D. B. (1986) The avian eye. In *Avian Physiology*, P. D. Stuckie, ed., pp. 39–50. Springer-Verlag, Berlin.

Meyer, R. (1977) Head capsule transmission of long-wavelength light in the *Circulimidae*. *Science*, **196**, 524–525.

Meyer-Rochon, V. B. (1973) The dioptric system in the beetle's compound eye. In *The Compound Eye and Vision of Insects*, ed. G. A. Horridge, pp. 299–310, Oxford Univ. Press, London.

Meyers, D. I. and Burger, M. M. (1977) Puzzling role of cell surfaces. *Chemistry*, **50**, 36.

Michelson, A. A. (1911) On the metallic colour of birds and insects. *Phil. Mag.*, **21**, 554.

Michelson, A. A. (1927) *Studies of Optics*. pp. 167–174, Univ. Chicago Press, Chicago.

Miller, G. (1983) *Modern Electronic Communications*. Prentice-Hall, Inc., Englewood Cliffs, New Jersey.

Miller, W. H. and G. D. Bernard (1968) Butterfly glow. *J. Ultrastruct. Res.*, **24**, 286.

Millot, N. (1968) The dermal light sense. In *Invertebrate Receptors*, J. D. Cathy and G. E. Newall, eds., pp. 1–36. Academic Press, New York.

Millot, N. (1978) *Extra-ocular photosensitivity*. Meadowfield Press, Bruham, England.

Mogus, M. A. and J. J. Wolken (1974) *Phycomyces:* electrical response to light stimuli. *Plant Physiol.*, **54**, 512.

Molyneaux, W. (1709) *Dioptrica Nova*, 2nd ed., printed for B. Tooke, London.

Moody, M. F. (1964) Photoreceptor organelles in animals. *Biol. Rev. Cambridge Phil. Soc.* **39**, 43.

Moore, T. (1953) Vitamin A in the normal individual. In *Symposium Nutrition*, R. M. Herriott, ed., p. 28, Johns Hopkins University Press, Baltimore.

Moore, W. (1989) *Schrödinger: Life and Thought*. Cambridge Univ. Press, New York.

Moray, N. (1972) Visual mechanisms in the copepod *Copilia*. *Perception*, **1**, 193.

Morison, W. L. (1984) Photoimmunology. *Photochem. Photobiol.*, **40**, 781–787.

Morton, R. A. (1944) Chemical aspects of the visual process. *Nature*, **153**, 69.

Morton, R. A. and T. W. Goodwin (1944) Preparation of retinene in vitro. *Nature*, **153**, 405.

Müller, H. (1851) Zur Histologie der Retina. *Z. Wiss. Zool.*, **3**, 234–237.

Müller, J. (1826) *Zur vergleichenden Physiologoe des Gesichtsinnes*. Cnobloch, Leipzig.

Munk, O. (1966) Ocular anatomy of some deep-sea teleosts, DANA. *Report Photobiol. 70*, pp. 60–62. Host & Son, Copenhagen.

Munk, O. (1980) *Hvirveldyrøjet; bygning, funktion og. tilpasning*. Berlingske, Copenhagen.

Munk, O. (1984) Non-spherical lenses in the eyes of deep-sea teleosts. *Arch. Fisch. Wiss.*, **34**, 145–153.

Muntz, W. R. A. (1987) Visual behavior and visual sensitivity of *Nautilus pompilius*. In *Living* Nautilus, N. H. Landewen and W. B. Sanders, eds., Plenum Press, New York.

Muntz, W. R. A. and U. Raj (1984) On the visual system of *Nautilus pompilius*. *J. Exp. Biol.*, **109**, 253–263.

Muntz, W. R. A. and S. L. Wentworth (1987) Anatomical study of the retina of *Nautilus pompilius*. *Biol. Bull.*, **173**, 387–397.

Munz, F. W. (1958) Photosensitive pigments from retinae of certain deep-sea fish. *J. Physiol.* **140**, 220–235.

Naka, K. I. (1960) Recording of retinal action potentials from single cells in the insect compound eye. *J. Gen. Physiol.*, **44**, 571.

Nathans, J. (1992) Rhodopsin structure, function and genetics. *Biochemistry*, **31**, 4923–4931.

Nathans, J., D. Thomas, and D. S. Hogness (1986a) Molecular genetics of human color vision: the genes encoding blue, green, and red pigments. *Science*, **232**, 193–202.

Nathans, J., T. P. Piantanida, R. L. Eddy, T. B. Shows, and D. S. Hogness (1986b) Molecular genetics of inherited variation in human color vision. *Science*, **232**, 203–210.

Needham, J. (1956) *Science and Civilization and China*, vol. 2, p. 298. Cambridge Univ. Press, London.

Neville, A. C. (1975) *Biology of Arthropod Cuticle*, Springer-Verlag, Berlin.

Neville, A. C. and S. Caveney (1969) Scarabaeid (June) beetle exocuticle as an optical analogue of cholesteric liquid crystals. *Biol. Rev. Cambridge Philos. Soc.*, **44**, 531.

Newton, I. (1666–1704) *Opticks: or, a Treatise of the Reflextions, Refractions, Inflecions and Colours of Light*, S. Smith & B. Walford, London. Modern reprints of the 1730 edition (corrected by the author). Republished by McGraw-Hill Book Co., New York, 1931, and by Dover Publications, New York, 1952.

Nilsson, D. E. (1988) A new type of imaging optics in compound eyes. *Nature*, **332**, 76.

Nilsson, D. E. (1989a) Vision optics and evolution. *Biol. Sci.*, **39**, 298–307.

Nilsson, D. E. (1989b) Optics and evolution of the compound eye. In *Facets of Vision*, D. G. Stavenga and R. Hardie, eds., Springer-Verlag, Berlin.

Nowikoff, M. (1932) Über den bau der komplexaugen von *Periplaneta (Stylopga) orientalis*. *Jena Z. Naturwiss*, **67**, 58.

O'Benar, J. D. and Y. Matsumoto, eds. (1976) Light-induced neural activity and muscle contraction in the marine worm, *Golfingia gouldii*. *Comp. Biochem. Physiol.*, **55A**, 77–81.

O'Brien, P. J. and D. C. Klein (1986) *Pineal and Retinal Relationship.* Academic Press, Orlando, FL.

Ogwa, T., Y. Inoue, M. Kitajima, and K. Shibata (1973) Action spectra for biosynthesis of chlorophyll a and b and β-carotene. *Photochem. Photobiol.,* **18,** 229.

Ohtsuka, T. (1978) Combination of oil droplets with different types of photoreceptors in a freshwater turtle (*Geoclemys reevesii*). *Sensory Proc.,* **2,** 321–325.

Ohtsuka, T. (1985) Relation of spectral types of oil droplets in cones of turtle retina. *Science,* **229,** 874–877.

Oldfield, E. (1973) Are cell membranes fluid? *Science,* **180,** 982.

Ootaki, T. and J. J. Wolken (1973) Octahedral crystals in *Phycomyces.* II. *J. Cell Biol.,* **57,** 278.

Palmer, J. (1974) *Biological Clocks in Marine Organisms,* Wiley, New York.

Park, R. B. and J. Biggins (1964) Quantasome: size and composition. *Science,* **144,** 1009.

Patten, W. (1887) Eyes of molluscs and arthropods. *J. Morphol.,* **1,** 67.

Pearse, J. S. and W. B. Pearse (1978) Vision in cubomedsan jellyfishes. *Science,* **199,** 458.

Peckman, G. W. and E. G. Peckman (1887) Some observations on the mental powers of spiders. *J. Morphol.,* **1,** 383.

Pelletier, J. and J. B. Caventou (1818) Sur la matière verte des feuilles. *Ann. Chim. Phys.,* **9,** 194.

Perralet, A. and F. Baumann (1969) Evidence for extracellular space in the rhabdom of the honeybee drone eye. *J. Cell. Biol.,* **3,** 825.

Pettigrew, G. W. and G. R. Moore (1987) *Cytochrome c: Biological Aspects,* Springer-Verlag, New York.

Planck, M. (1922) *The Origin and Development of the Quantum Theory,* trans. H. T. Clarke and L. Silberstein, Oxford Univ. Press, Oxford.

Pollock, J. A. and S. Benzer (1988) Transcript localization of four opsin genes in three visual organs of *Drosophila;* RH2 is ocellus specific. *Nature,* **333,** 779–783.

Polyak, S. L. (1957) *The Vertebrate Visual System,* H. Klüver, ed., Univ. Chicago Press, Chicago.

Presti, D. E. and P. Galland (1987) *Photoreceptor Biology of Phycomyces,* E. Cerdá-Olmedo and E. D. Lipson, eds., pp. 93–126. Cold Spring Harbor Laboratory. Cold Spring Harbor, NY.

Priestly, J. (1772) Observations on different kinds of air. *R. Soc. London Phil. Trans.,* **62,** 147.

Pugh, E. N., Jr. and W. H. Cobbs (1986) Visual transduction in vertebrate rods and cones: a tale of two transmitters, calcium and cyclic GMP. *Vis. Res.* **26,** 1613–1643.

Pumphrey, R. J. (1961) *Concerning Vision, in the Cell and the Organism,* J. A. Ramsay and V. B. Wigglesworth, eds., Cambridge Univ. Press, Cambridge.

Purkinje, J. (1825) *Beobachtungen und Versuche zur Physiologie der Sinne,* Vol. 2, Reimer, Berlin.

Quay, W. B. (1986) Indole biochemistry and retinal mechanisms. In *The Pineal and Retinal Relationship,* P. J. O'Brien and D. C. Klein, eds., pp. 107–118, New York.

Rabinowitch, E., (1945, 1951, 1956) *Photosynthesis and Related Processes,* 3 Vol., Wiley Interscience, New York.

Rayport, S. and G. Wald (1978) Frog skin photoreceptors, *Am. Soc. Photobiol. Abstr.,* **6,** 94–95.

Regan, J. D. and J. S. Cooke (1969) Photoreactivation in an established vertebrate cell line. *Proc. Natl. Acad. Sci. U.S.A.,* **58,** 2274.

Robertson, J. D. (1959) The ultrastructure of cell membranes and their derivatives. *Biol. Chem. Symp.*, **16**, 3–43.

Robinson, C. (1966) The cholesteric phase in polypeptides and biological structures. *Mol. Cryst.*, **1**, 467.

Rochon-Duvigneaud, A. (1943) *Les Yeux et la Vision des Vertebraes*, Masson, Paris.

Röhlich, P. (1966) Sensitivity of regenerating and degenerating planarian photoreceptors to osmium tetraoxide. *Z. Zellforsch. Mikrosk. Anat.*, **73**, 165.

Rölich, P., and I. Törö, (1965) Fine structure of the compound eye of *Daphnia* in normal, dark-adapted and strongly light-adapted states. In *The Eye Structure II Symposium.* J. W. Rohen and F. K. Schattauer-Verlag, Stuttgart. pp. 175–186.

Röhlich, P. and I. Törö (1961) Electronenmikroscopische untersuchungen des Auges von Planarien. *Z. Zellforsch. Mikrosk. Anat.*, **54**, 362.

Romer, A. S. (1955) *The Vertebrate Body,* pp. 34-35, W. B. Saunders Company, Philadelphia (2nd ed.).

Romer, G. B. and J. Delamoir (1989) The first color photographs. *Sci. Am.*, **261**, 88–96.

Rosenberg, A. (1967) Galactosyl diglycerides: their possible function in *Euglena* chloroplasts. *Science*, **157**, 1191.

Ross, E. M. (1988) Receptor-G protein-effector. The design of a biochemical switchboard. *J. Chem. Ed.*, **66**, 937–942.

Rossel, S. and R. Wehner (1986) Polarization vision in bees. *Nature*, **323**, 128–131.

Sager, R. (1972) *Cytoplasmic Gene and Organelles,* Academic Press, New York.

Scharrer, E. (1964) A specialized trophosporangium in large neurons of *Leptodora* (Crustaceae). *Z. Zellforsch. Mikrosk. Anat.*, **61**, 803.

Schmidt, W. (1984) Bluelight physiology. *Bioscience*, **34**, 698.

Schrödinger, E. (1928) *Collected Papers on Wave Mechanics* (translated from 2nd German Ed.), Blachie and Son Ltd., London, England.

Schultze, M. J. (1866) Zur anatomie und physiologie der retina. *Arch. Mikrosk. Anat.*, **2**, 175.

Schwind, R. (1989) *Size and Distance Perception in Compound Eyes in Facets of Vision,* D. C. Stavenga and R. C. Hardie, eds., Springer-Verlag, New York.

Senebier, J. (1782) *Memoires physico-chimiques sur l'influence de la lumière solaire pour modifier les etros de trois regnes, sûrtout ceux de regne végétal,* 3 vols., Chirol, Geneva.

Senger, H., ed. (1987) *Blue Light Responses,* vols. 1 and 2, CRC Press, Boca Raton, FL.

Serebrovakaya, I. (1971) *Chemical Evolution and the Origin of Life,* R. Buvet and C. Ponnamperuma, eds., pp. 297–306, Elsevier, New York.

Setlow, J. E. and R. B. Setlow (1963) Nature of the photoreactivable lesion in deoxyribonucleic acid (DNA). *Nature*, **197**, 560.

Shemin, D. (1948) The biosynthesis of porphyrins. *Cold Spring Harbor Symp. Quant. Biol.*, **13**, 185–192.

Shemin, D. (1955) The succinate-glycine cycle: the role of delta-amino-levulinic acid in porphyrin synthesis. *Porphyrin Biosyn. Metab.*, *Ciba Found. Symp.*, 4–28.

Shemin, D. (1956) The biosynthesis of porphyrins. In *Harvey Lect.*, 1954–1955. Series L. **50**, 258–284.

Shropshire, W., Jr. (1963) Photoresponses of the fungus *Phycomyces. Physiol. Rev.*, **43**, 38.

Shurcliff, W. A. (1962) *Polarized Light: Production and Use,* pp. 43–50, 149–151, Harvard Univ. Press, Cambridge, MA.

Shurcliff, W. A. (1955) Haidinger's brushes and circularly polarized light, *J. Opt. Soc. Am.* **45**, 399.

Singer, S. J. (1971) The molecular organization of biological membranes. In *Structure and Function of Biological Membranes*, L. I. Rothfield, ed. pp. 145–222, Academic Press, New York.

Singer, S. J., and G. L. Nicholson (1972) The fluid mosaic model of the structure of cell membranes. *Science,* **175**, 720.

Sivak, J. G. (1980) Vertebrate strategies for vision in air and water. In *Sensory Ecology,* M. A. Ali, ed., pp. 503–519, Plenum Press, New York.

Smith, A. M. (1987) Descarte's theory of light and refraction. *Trans. Am. Phil. Society.* **77**, pt. 3.

Snyder, A. W. and W. H. Miller (1978) Telephoto lens system of falconiform eyes. *Nature,* **275**, 127.

Sominya, H. (1976) Functional significance of the yellow lens in the eyes of *Arsyropelecus affinis. Marine Biol.,* **34**, 93–99.

Stanier, R. Y. (1959) Formation and function of the photosynthetic pigment in purple bacteria. *Brookhaven Symp. Biol.,* **11**, 43–53.

Stavenga, D. G. and R. C. Hardie, eds., (1989) *Facets of Vision.* Springer-Verlag, Berlin.

Stephens, G. C., M. Fingerman, and F. A. Brown (1952) A nonbirefringent mechanism for orientation to polarized light in Arthropods. *Anat. Rec.,* **113**, 559–560.

Steven, D. M. (1963) The dermal light sense. *Cambridge Phil. Soc. Biol. Rev.,* **38**, 204–240.

Stevens, J. K. and K. E. Parsons (1980) A fish with double vision. *Nat. Hist.,* **89**, 62–67.

Stoeckenius, W. R. H. Lozier, and R. H. Bogomolni (1979) Bacteriorhodopsin and the purple membrane, *Halobacteria. Biochem. Biophys. Acta,* **505**, 215–278.

Strother, G. K. (1963) Absorption spectra of retinal oil globules in turkey, turtle, and pigeon. *Exp. Cell Res.,* **59**, 249.

Strother, G. K. (1977) *Physics: With Applications in Life Sciences.* Houghton Mifflin, Boston.

Strother, G. K. and A. J. Casella (1972) Microspectrophotometry of arthropod visual screening pigments. *J. Gen. Physiol.,* **59**, 616.

Strother, G. K. and J. J. Wolken (1960) Microspectrophotometry. I. Absorption spectra of colored oil globules in the chicken retina. *Exp. Cell Res.* **21**, 504.

Stuermer, W. (1970) Soft parts of cephalopods in trilobites: some surprising results of X-ray examinations of Devonian slates. *Science,* **170**, 1300.

Sutherland, B. C. and B. M. Sutherland (1975) Human photoreactivity enzyme: Action spectrum and safelight condition. *Biophys.J.,* **15**, 435–440.

Sutherland, B. M., H. L. Barber, L. C. Harber, and I. Kochevar (1980) Pyrimidine dimer formation and repair in human skin. *Cancer Res.,* **40**, 3181–3185.

Tamarkin, L., C. J. Baird, and O. F. Almeida (1985) Melatonin: coordinating signal for mammalian reproduction. *Science,* **227**, 714–727.

Tanford, C. (1980) *The Hydrophobic Effect: Formation of Micelles and Biological Membranes,* 2nd ed. p. 233, Wiley, New York.

Taylor, D. H. (1972) Extra-optic photoreception and compass orientation in larval and adult salamanders, *Cambystoma tigrinum. Anim. Behav.,* **20**, 233–236.

Taylor, D. H. and D. E. Ferguson (1970) Extraoptic celestial orientation in the southern cricket frog, *Acirs gryllus. Science,* **168**, 390–392.

Tokioka, R., K. Matsuoka, Y. Nakaoka, and Y. Kito (1991) Extraction of retinal from *Paramecium bursaria*. *Photochem. Photobiol.*, **53**, 149–151.

Tomita, T. (1970) Electrical activity of vertebrate photoreceptors. *Q. Rev. Biophys.*, **3**, 179.

Towe, K. M. (1973) Trilobite eyes calcified lenses *in-vivo*. *Science*, **179**, 1007.

Truman, J. W. (1976) Extra retinal photoreception in insects. *Photochem. Photobiol.*, **23**, 215.

Tsuda, M. (1987) Photoreception and phototransduction in invertebrate photoreceptors. *Photochem. Photobiol.*, **45**, 915–931.

Underwood, H. (1977) Circadian organization in lizards: the role of the pineal organ. *Science*, **195**, 587–589.

Unwin, P. N. T. and R. Henderson (1975) Molecular structure determination by electron microscopy of unstained crystalline specimens. *J. Mol. Biol.*, **94**, 425.

Vaissière, R. (1961) *Morphologie et histologie comparées des yeux des crustacés copépodes*. Thesis, Centre National de la Recherche Scientifique, Paris.

Van Brunt, E. E., M. D. Shepard, J. R. Wall, W. F. Ganong, and M. T. Clegg (1964) Penetration of light into the brain of mammals. *Ann. N.Y. Acad. Sci.*, **117**, 204–216.

Vanderkooi, G. and D. E. Green (1971) New insights into biological membrane structure. *BioScience*, **21**, 409.

van Dorp, D. A. and J. F. Arens (1947) Synthesis of vitamin A aldehyde. *Nature*, **160**, 189.

van Leeuwenhoek, A. (1674) *Phil. Trans. R. Soc. Lond.*, **9**, 198. More observations from Mr. Leeuwenhoek. (see Dobell, 1958.)

Van Niel, C. B. (1941) The bacterial photosyntheses and their importance for the general problem of photosynthesis. *Advan. Enzymol.*, **1**, 263.

Van Niel, C. B. (1943) Biochemical problems of the chemo-autotrophic bacteria. *Physiol. Rev.*, **23**, 338.

Van Niel, C. B. (1949) The comparative biochemistry of photosynthesis. *Am. Sci.*, **37**, 371.

Van Veen, T., H. G. Hartwig, and K. Muller (1976) Light-dependent motor activity and photonegative behavior in the eel *Anguilla anguilla:* evidence for extra retinal photoreception. *J. Comp. Physiol.*, **111A**, 209–219.

Varela, F. G. and W. Wiitanen (1970) The optics of the compound eye of the honeybee *(Apis mellifera). J. Gen. Physiol.*, **55**, 336.

Vogt, K. (1977) Ray path and reflection mechanism of crayfish eyes. *Z. Naturforsch.*, **32C**, 466–468.

Vogt, K. (1980) Die Spiegel optick des flußkrebsauges. *J. Comp. Physiol.*, **135**, 1–19.

von Frisch, Karl (1914) Der farbenensinn und formensinn der beinen. *Zool. Jahrb., Abt. Allg. Zool. Physiol. Tiere*, **35**, 1.

von Frisch, Karl (1949) Die polarisation des himmelslichtes als faktor der orientieren bei den tänzen der bienen. *Experientia*, **5**, 397.

von Frisch, Karl (1950) *Bees: Their Vision, Chemical Senses, and Language*. Cornell Univ. Press, Ithaca, NY (rev. ed. 1971).

von Frisch, Karl (1967) *The Dance Language and Orientation of Bees*. Belknap Press, Harvard Univ., Cambridge, MA.

von Helmholtz, H. (1852) Über die Theorie der zusammangestzten Farben. *Ann. Physz. (Leipzig)*, **87**, 45.

von Helmholtz, H. (1867) *Handbuch der Physiologischen Optic*. Voss, Leipzig. (Republished *Physiological Optics,* edited by J. P. C. Southall, vols. I, II, III, 1924–25. Optical Society of Amer.: Rochester, NY. and Dover Publication, Inc. New York.)

von Sachs, J. (1864) Wirkungen Farbigen Lichts auf Pflanzen. *Botan. Z.,* **22,** 353.

Vowles, D. M. (1954) The orientation of ants. II. Orientation to light, gravity, and polarized light. *J. Exp. Biol.,* **31,** 356.

Wald, G. (1933) Vitamin A in the retina. *Nature,* **132,** 316–17.

Wald, G. (1935) Carotenoids and the visual cycle. *J. Gen. Physiol.,* **19,** 351–371.

Wald, G. (1948) Galloxanthin, a carotenoid from the chicken retina. *J. Gen. Physiol.,* **31,** 377.

Wald, G. (1952) Biochemical evolution. In *Modern Trends in Physiology and Biochemistry,* E. S. G. Barrón, ed., pp. 337–376, Academic Press, New York.

Wald, G. (1953) The biochemistry of vision. *Annu. Rev. Biol.,* **22,** 497.

Wald, G. (1954) On the mechanism of the visual threshold and visual adaptation. *Science,* **119,** 887.

Wald, G. (1956) The biochemistry of visual excitation. In *Enzymes: Units of Biological Structure and Function,* O. H. Gaebler, ed., p. 355, Academic Press, New York.

Wald, G. (1959) The photoreceptor process in vision. In *Handbook of Physiology,* J. Field, ed., sect. 1, vol. 1, p. 671, Williams and Wilkins (Am. Physiol. Soc.), Baltimore.

Wald, G. (1964) General discussion of retinal structure in relation to the visual process. In *Structure of the Eye,* G. K. Smelser, ed., pp. 101–115, Academic Press, New York.

Wald, G. (1970) Vision and the mansions of life, the first Feodor Lynen Lecture in "Miami Winter Symposia," pp. 1–32, North-Holland, Amsterdam.

Wald, G. (1973) Visual pigments and photoreceptor physiology. In *Biochemistry and Physiology of Visual Pigments,* pp. 1–13, Springer-Verlag, New York.

Wald, G. and P. K. Brown (1965) Human color vision and color blindness. *Cold Spring Harbor Symp. Quant. Biol.,* **30,** 345.

Wald, G. and S. Rayport (1977) Vision in annelid worms. *Science,* **196,** 1434–1439.

Wald, G. and H. Zussman (1938) Carotenoids of the chicken retina. *J. Biol. Chem.,* **142,** 449.

Waldvogel, J. A. (1990) The birds' eye view. *Am. Sci.,* **78,** 342–353.

Walls, G. (1942) *The Vertebrate Eye,* Cranbrook Institute of Science, Bloomfield Hills, MI.

Warburg, O. and W. Christian (1938a) Coenzyme of the d-amino acid deaminase. *Biochem. Z.,* **295,** 261.

Warburg, O. and W. Christian (1938b) Coenzyme of the d-alanine-dehydrogenase. *Nat. Wiss.,* **26,** 235.

Warburg, O. and W. Christian (1938c) Coenzyme of d-alanine oxidase. *Biochem. Z.,* **296,** 294.

Watanabe, M. (1980) An autoradiographic, biochemical and morphological study of the Harderian gland of the mouse. *J. Morphol.,* **163,** 349–365.

Waterman, T. H. (1975) The optics of polarization sensitivity. In *Photoreceptor Optics,* A. W. Snyder and R. Menzel, eds., pp. 319–370, Springer, Berlin.

Waterman, T. H. (1984) Natural polarized light and vision. In *Photoreception and Vision in Invertebrates,* M. A. Ali, ed., pp. 63–114. Plenum, New York.

Waterman, T. H. (1989) *Animal Navigation,* W. H. Freeman and Company, New York.

Waterman, T. H., H. R. Fernandez, and T. H. Goldsmith (1969) Dichroism of photosensitive pigment in rhabdoms of the crayfish *Orconectes. J. Gen. Physiol.,* **54,** 415.

Watson, L. (1973) *Super Nature.* Anchor Press Doubleday, Garden City, NY.

Weiderhold, M. L., E. F. MacNichol, Jr., and A. L. Bell (1973) Photoreceptor spike response in the hardshell clam, *Mercenaria mercenaria. J. Gen. Physiol.*, **61**, 24–55.

Welford, W. T. and R. Winston (1989) *High Collection Non-Imaging Optics.* Academic Press, Orlando, FL.

Wells, M. J. (1978) *Octopus: Physiology and Behavior of an Advanced Invertebrate,* Chapman and Hall, London.

Welsh, J. H. (1964) The quantitative distribution of 5-hydroxytryptamine in the nervous system, eyes, and other organs of some vertebrates. In *Proceedings of the International Neurochemistry Symposium* (5th), C. D. Richert, ed., pp. 355–366, St. Wolfgang, Austria.

Wetterberg, L., A. Uywiler, E. Geller, and S. Schapiro (1970a) Harderian gland: development and influence of early hormonal treatment of porphyria content. *Science*, **168**, 996–998.

Wetterberg, L., E. Geller, and A. Yuwiler (1970b) Harderian gland: an extraretinal photoreceptor influencing the pineal gland in the neonatal rat. *Science*, **171**, 194–196.

Wetterberg, L., A. Yuwiler, R. Ulrich, E. Geller, and R. Wallace (1970c) Harderian gland: influence on pineal hydroxyindole-O-methyltransferase activity in neonatal rats. *Science*, **170**, 194–196.

Wiener, N. (1964) Problems of sensory prosthesis. In *Selected Papers of Norbert Wiener,* pp. 431–439, MIT Press, Cambridge, MA.

Wigglesworth, V. B. (1964) *The Life of Insects,* Wiedenfield & Nicholson, London.

Wilkens, L. A. and J. S. Larimer (1972) The CNS photoreceptor of crayfish: morphology and synaptic activity. *J. Comp. Biol.*, **80**, 389–407.

Wilkens, L. A. and J. J. Wolken (1981) Electroretinograms from *Odontosyllis enopla* (Polychaete: Sylladae): initial observations on the visual system of the bioluminescent fireworm of Bermuda. *Marine Behav. Physiol.*, **8**, 55–56.

Willmer, E. N., ed. (1960) *Cytology and Evolution,* 1st ed. Academic Press, New York.

Willstäter, R. and A. Stoll (1913) *Untersuchungen uber Chlorophyll. Methoden und Ergebnisse,* Springer-Verlag, Berlin.

Winston, R. (1975) Light collection within the framework of geometrical optics. *J. Opt. Soc. Am.*, **60**, 245.

Witman, C. G., K. Carlson, J. Berliner, and J. L. Rosenbaum (1972a) *Chlamydomonas* flagella. I. Isolation and electrophoretic analysis of microtubules, matrix, membranes, and mastigonemes. *J. Cell Biol.*, **54**, 507.

Witman, C. G., K. Carlson, and J. L. Rosenbaum (1972b) *Chlamydomonas* flagella. II. The distribution of tubulins 1 and 2 in the outer doublet microtubule. *J. Cell Biol.*, **54**, 540.

Wolken, J. J. (1966) *Vision: Biochemistry and Biophysics of the Retina,* Thomas, Springfield, IL.

Wolken, J. J. (1967) *Euglena: An experimental organism for biochemical and biophysical studies,* 2nd ed., Appleton, New York.

Wolken, J. J. (1971) *Invertebrate Photoreceptors: A Comparative Analysis,* Academic Press, New York.

Wolken, J. J. (1972) *Phycomyces*, a model neurosensory cell. *Int. J. Neurosci.* **3**, 135.

Wolken, J. J. (1975) *Photoprocesses, Photoreceptors, and Evolution,* Academic Press, New York.

Wolken, J. J. (1977) *Euglena*, the photoreceptor system for phototaxis. *J. Protozool.*, **24**, 518–522.

Wolken, J. J. (1984) Self-organizing molecular systems. In *Molecular Evolution and Photo-biology,* K. Matsuno, K. Dose, K. Harada, and D. L. Rholfing, eds., pp. 137–162, Plenum Press, New York.

Wolken, J. J. (1986) *Light and Life Processes.* Van Nostrand Reinhold, New York.

Wolken, J. J. (1987) *Light Concentration Lens System.* United States Patent Office numbers 4,669,532 and 4,988,177 (1991).

Wolken, J. J. (1988) Photobehavior of marine invertebrates: Extraocular photoreception. *Comp. Biochem. Physiol., 91,* 145–149.

Wolken, J. J. (1989) Biomimetics: light detectors and imaging systems in nature. In *Bio-technology for Aerospace Application.* I. W. Obringerand and T. S. Tellinghast, eds., pp. 157–164, Gulf Publishing Company, Houston, TX.

Wolken, J. J. and R. G. Florida (1969) The eye structure and optical system of the crustacean copepod, *Copelia. J. Cell. Biol., 40,* 279.

Wolken, J. J. and R. G. Florida (1984) The eye structure of the bioluminescent fireworm of Bermuda, *Odontosyllis enopla. Biol. Bull, 166,* 260–268.

Wolken, J. J. and G. J. Gallik (1965) The compound eye of the crustacean: *Leptodora kindtii. J. Cell Biol., 26,* 968–973.

Wolken, J. J. and J. A. Gross (1963) Development and characteristics of the *Euglena* c-type cytochromes. *J. Protozool., 10,* 189.

Wolken, J. J. and P. D. Gupta (1961) Photoreceptor structures of the retinal cells of the cockroach eye. IV *Periplaneta americana* and *Blaberus giganteus. J. Biophys. Biochem. Cytol., 9,* 720.

Wolken, J. J. and M. A. Mogus (1979) Extra-ocular photosensitivity. *Photochem. Photo-biol., 29,* 189–196.

Wolken, J. J. and M. A. Mogus (1981) Extraocular photoreception. In *Photochemistry and Photobiological Review,* K. Smith, ed., Vol. 6, pp. 181–189, Plenum Press, New York.

Wolken, J. J. and M. A. Mogus (1988) A light concentrating lens. *Penna. Acad. Sci., 62,* 100.

Wolken, J. J. and C. S. Nakagawa (1973) Rhodopsin formed from bacterial retinal and cattle opsin. *Biochem. Biophys. Res. Commun., 54,* 1262.

Wolken, J. J., J. Capenos, and A. Turano (1957) Photoreceptor structures. *J. Biophys. Biochem. Cytol., 3,* 441–448.

Wood, R. W. *Physical Optics,* Republished from original 1911 edition by Opt. Soc. Am., Washington, D.C., 3rd ed. 1988.

Woodward, R. B. (1961) Total synthesis of chlorophyll. *Pure Appl. Chem., 2,* 383.

Wurtman, R. J. (1975) The effects of light on the human body. *Sci. Am., 233,* 68–77.

Wurtman, R. J., and J. Axelrod (1965) The Pineal gland, *Sci. Am., 213,* 50–60.

Wurtman, R. J., J. Axelrod and D. E. Kelly (1968) *The Pineal.* Academic Press, New York.

Wurtman, R. J., M. J. Baum, and J. T. Potts, Jr., eds. (1985) The medical and biological effects of light. *Ann. N.Y. Acad. Sci., 453.*

Yoshida, M. (1979) Extraocular photoreception. In *Handbook of Sensory Physiology,* Vol 7/6A, *Vision in inverterates, A: Invertebrate photoreceptors,* H. Autrum, ed., pp. 581–640, Springer-Verlag, Berlin.

Yoshida, M., H. Oshtsuki, and S. Siguri (1967) Ommochrome from anthomedusan ocelli and its photoreduction. *Photochem. Photobiol., 6,* 875.

Yoshizawa, T. and O. Kuwata (1991) Iodopsin, a red-sensitive cone visual pigment in the chicken retina. *Photochem. Photobiol., 54,* 1061–1070.

Young, J. Z. (1964) *A Model of the Brain*. Oxford Univ. Press, New York.

Young, J. Z. (1971) *Introduction to the Study of Man*. Oxford Univ. Press, New York.

Young, T. (1802) On the theory of light and colors. *R. Soc. London Phil. Trans.*, **92,** 12.

Young, T. (1803) Experiments and calculations relative to physical optics. *Phil. Trans. R. Soc. Lond.*, **94,** 1, (1804). A course of lectures on natural philosophy, Vol. 2, p. 639. German trans. in *Ann. D. Physik,* Vol. 39, p. 262 (1811).

Young, T. (1807) *Lectures on Natural Philosophy,* vol. 1., pp. 315, 613. W. Savage, London.

Zacks, D. N., F. Derguini, K. Nakanishi, and J. L. Spudich (1993) Comparative study of phototactic and photophobic receptor chromophore properties in *Chlamydomonas reinhardtii, Biophys. J.,* **65,** 508–518.

Zechmeister (1962) *Carotenoids: Cis-Trans Isometric Cartenoids, Vitamin A and Arylpolyenes,* Academic Press, New York.

Zeigler, H. P. and H.-J. Bischof, eds. (1993) *Vision, Brain and Behavior in Birds*. M.I.T. Press, Cambridge, Mass.

Ziegler, I. (1964) Über naturlich vorkommende Tetahydropropterine. In *Pteridine Chemistry,* W. Pfeiderer and E. C. Taylor, eds., pp. 295–305, Pergamon, Oxford.

Ziegler, I. (1965) Pterine als Wirkstoffe und Pigmente. *Ergeb. Physiol. Biol. Chem. Exp. Pharmakol.,* **56,** 1.

Ziegler-Gunder, I. (1956) Pterine: Pigmente und Wirkstoffe im Tierreich. *Biol. Rev. Cambridge Phil. Soc.,* **31,** 313.

Zigman, S. (1991) Comparative biochemistry and biophysics of Elasmobranch lenses. *J. Exp. Zool.,* **Supp. 5,** 29–40.

Zinter, J. R. (1987) *A Three-Dimensional Superposition Array,* Ph.D. Thesis. Univ. Rochester.

Zonana, H. V. (1961) Fine structure of the squid retina. *Bull. Johns Hopkins Hosp.,* **109,** 185.

Zrenner, C. (1985) Theories of pineal function from classical antiquities to 1900. In *Pineal Research Reviews,* R. J. Reiter, ed., vol. 3, pp. 1–40, Alan R. Liss Inc., New York.

Zweibel, K. (1990) *Harnessing Solar Power,* **6,** Plenum Press, New York.

Index